Drilling Technology
in nontechnical language

Drilling Technology
in nontechnical language

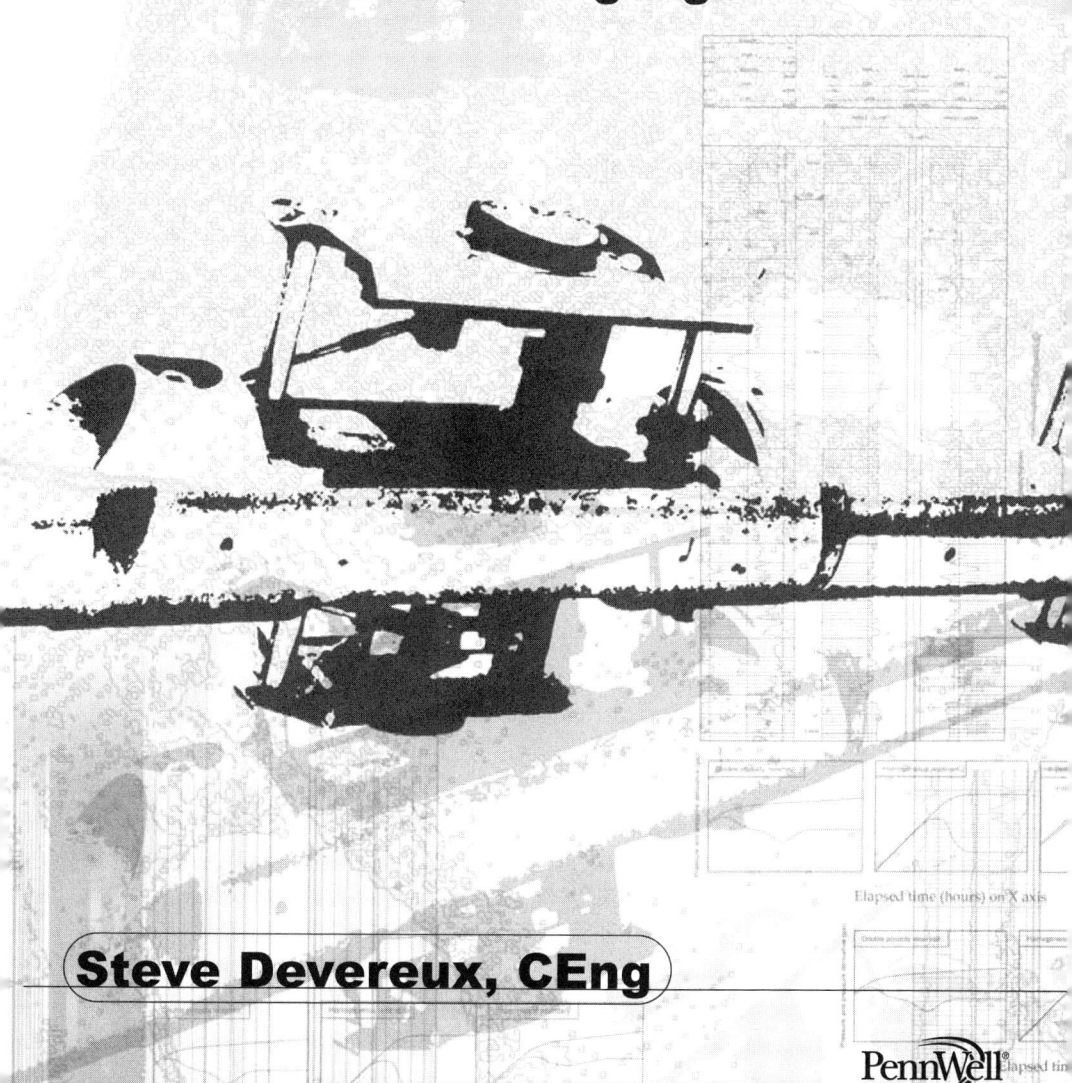

Steve Devereux, CEng

PennWell

This book is dedicated to my family —
my wife, Trish,
my daughter, Robyn, and my sons, Alex and Michael.

Disclaimer: The recommendations, advice, descriptions, and the methods in this book are presented solely for educational purposes. The author and publisher assume no liability whatsoever for any loss or damage that results from the use of any of the material in this book. Use of the material in this book is solely at the risk of the user.

Copyright © 1999 by
PennWell Corporation
1421 South Sheridan Road
Tulsa, Oklahoma 74112-6600 USA

800.752.9764
+1.918.831.9421
sales@pennwell.com
www.pennwellbooks.com
www.pennwell.com

Marketing Manager: Julie Simmons
National Account Executive: Barbara McGee
Director: Mary McGee
Managing Editor: Marla Patterson
Production Manager: Sheila Brock

Library of Congress Cataloging-in-Publication Data
Devereux, Steve
Drilling technology in nontechnical language / Steve Devereux
p.cm.
Includes bibliographical references and index
ISBN 0-87814-672-4
ISBN13 978-0-87814-762-5
1. Oil well drilling. I. Title.
TN871.2.D477 1999
622'.3381--dc21   99-055192

All rights reserved. No part of this book may be reproduced, stored in a retrieval system, or transcribed in any form or by any means, electronic or mechanical, including photocopying and recording, without the prior written permission of the publisher.

Printed in the United States of America

6 7 8 9 10   11 10 09 08 07

Table of Contents

Figures/Tables List xiii
Introduction xvii

Chapter 1 Drilling Geology 1
Chapter Overview 1
Igneous, Metamorphic, and Sedimentary Rocks 1
Plate Tectonics 3
Lithology 3
Rock Strengths and Stresses 8
Hydrostatic Pressure Imposed by a Fluid 9
Chapter Summary 12
Glossary 12

Chapter 2 Oil and Gas Generation, Migration, and Reservoirs 15
Chapter Overview 15
Source Rock and Hydrocarbon Generation 16
Vital Rock Properties 17
Primary Migration 19
Structural Trap 20
Reservoir Rock 20
Seal Rock 21
Secondary Migration 22
Reservoir Drives 23
Problems Related to Fluids in the Reservoir 24
Chapter Summary 25
Glossary 26

Chapter 3	Planning and Drilling an Exploration Well on Land	27
	Chapter Overview	27
	Identifying a Prospect	27
	Well Proposal	28
	Gathering Data	30
	Designing the Well	33
	Writing the Well Program	36
	Drilling the Well	44
	Production Testing the Well	70
	Abandoning the Well	73
	Chapter Summary	74
	Glossary	74
Chapter 4	Planning and Drilling a Development Well Offshore	79
	Chapter Overview	79
	Well Planning	79
	Hole and Casing Sizes	82
	Writing the Well Program	85
	Drilling the Well	86
	Chapter Summary	100
	Glossary	100
Chapter 5	Rig Selection and Rig Equipment	103
	Chapter Overview	103
	Selecting a Suitable Drilling Rig	103
	Classifications of Drilling Rigs—Description	104
	Rig Systems and Equipment	113
	Chapter Summary	129
	Glossary	129

Table of Contents

Chapter 6	Drill Bits 133
	Chapter Overview 133
	Roller Cone Bits..134
	Fixed Cutter Bits 135
	Core Bits 137
	Optimizing Drilling Parameters 138
	Grading the Dull Bit 141
	Bit Selection...142
	Drill Bit Economics 143
	Chapter Summary 144
	Glossary...144

Chapter 7	Drilling Fluids 145
	Chapter Overview 145
	Functions of the Drilling Fluid 145
	Basic Mud Classifications...146
	Designing the Drilling Fluid 151
	Chapter Summary 167
	Glossary...167

Chapter 8	Directional and Horizontal Drilling 171
	Chapter Overview 171
	Why Drill Directional Wells? 171
	Tools and Techniques for Kicking Off the Well 174
	Controlling the Wellpath of a Deviated Well 178
	Horizontal Wells 181
	Multilateral Wells...183
	Surveying...184
	Navigation by Reference to Reservoir Characteristics......187
	Chapter Summary 188
	Glossary...188

Drilling Technology in Nontechnical Language

Chapter 9	Casing and Cementing	189
	Chapter Overview	189
	Importance of Casing in a Well	190
	Designing the Casing String	192
	Role of the Cement Outside Casing	198
	Mud Removal	198
	Cement	200
	Cement Design	201
	Running and Cementing Casing	208
	Cementing Surface Casings	210
	Cement Evaluation Behind Casing	211
	Other Cement Jobs	211
	Chapter Summary	213
	Glossary	214
Chapter 10	Evaluation	215
	Chapter Overview	215
	Evaluation Techniques	215
	Physical Sampling at Surface	216
	Physical Sampling Downhole	220
	Electrical Logging	225
	Production Testing	232
	Chapter Summary	235
	Glossary	235
Chapter 11	Well Control	237
	Chapter Overview	237
	Primary, Secondary, and Tertiary Well Control	237
	BOP Stack	239

Table of Contents

	Kick Detection Equipment	247
	Killing the Well	249
	Shallow Gas	254
	Special Well Control Considerations	256
	Certification of Personnel for Well Control	261
	Chapter Summary	261
	Glossary	261
Chapter 12	Managing Drilling Operations	263
	Chapter Overview	263
	Personnel Involved in Drilling Operations	263
	Contract Types	268
	Incentive Schemes	270
	Decision Making at the Wellsite	272
	Decision Making in the Office	272
	Interfacing with Service Companies	273
	Estimating the Well Cost	274
	Logistics	277
	Handling Major Incidents	279
	Chapter Summary	280
	Glossary	281
Chapter 13	Drilling Problems and Solutions	283
	Chapter Overview	283
	Lost Circulation	283
	Stuck Pipe	290
	Fishing	298
	Chapter Summary	306

Chapter 14 Safety and Environmental Issues 307
 Chapter Overview 307
 Safety Meetings 307
 Newcomers on the Rig..310
 Training and Certification 311
 Drills 312
 Permit to Work Systems 313
 Safety Alerts...314
 Equipment Certification..314
 Safety Equipment...316
 STOP 316
 Minimizing Discharge and Spills 317
 Environmental Impact Studies......................................319
 Severe Weather—Suspension of Operations 320
 Chapter Summary 320
 Glossary...320
Index...323

Figures

Fig.	Heading	Page #
1–1	Bedding Formations	7
1–2	Comparison of Compressive Strengths	9
1–3	Square Liquid Storage Tube	10
2–1	Typical Sandstone Structure	18
2–2	Stratigraphic Traps	20
3–1	Illustration of a Packer	31
3–2	Bit Drilling Through Dipping Beds	32
3–3	Sequence of Drilling a Well and Running Casings	34
3–4	Diagram of Shallow Gas	40
3–5	Time—Depth for Well Example 1	41
3–6	Diagram of a Drilling Rig	47
3–7	Drawing Showing How the Diverter Is Set Up On the Conductor Before Drilling Starts	49
3–8	Illustration of the Use of Solids-control Equipment	51
3–9	A Bit Sub	53
3–10	Example of a Drilling Tally Sheet	54
3–11	A Hole Opener and Bullnose	56
3–12	Situation at the End of the Surface Casing Cement Job	57
3–13	Intermediate Casing Suspended In the Casing Head Housing	58
3–14	Use of BOP Equipment	59
3–15	Stabilizer—Conceptual Drawing	62
3–16	Intermediate Casing Before Landing In the Casinghead Housing	65
3–17	Intermediate Casing Landed and Ready to Start Cementing	66
3–18	Building Up the Wellhead for the Next Hole Section	67
3–19	Final Well Profile With the Production Casing and Liner In Place	69
3–20	Model of a PDC Core Bit Seen from Below	70
3–21	Wellhead Before Nippling Up the Christmas Tree	73
4–1	Directional Wells Drilled from a Floating Rig Through a Template	80
4–2	A Sand Control Screen, Cut Away to Show the Screen Elements	81

4–3	Coning—Gas Breaking Through Oil	82
4–4	Traditional Vs. Monobore Completion	83
4–5	Directional Well Terminology	84
4–6	Template Set On the Seabed (side view)	86
4–7	Wellhead Housing Welded to Surface Casing	88
4–8	BOP Latched Onto Wellhead Housing	89
4–9	Arrangement of Telescopic Joint and Tensioners	90
4–10	Photograph of the Telescopic Joint	91
4–11	Heave Compensator—Principle of Operation	92
4–12	Calculating Depths In a Deviated Well	95
4–13	4 Arm Caliper, Used to Measure Hole Diameter In Two Axes	97
4–14	Identifying the Second Kickoff Point	98
5–1	Drilling Rig Classification	104
5–2	Land Rig In the Desert	105
5–3	Semi-submersible Rig With Supply Boat	107
5–4	Drilling Tender With Derrick On Platform	108
5–5	Drilling Barge With Derrick Stowed for Moving	109
5–6	Jackup Rig, Under Tow and On Location	110
5–7	Jackup Rig Drilling Over a Platform	111
5–8	Conventional Platform With Self Contained Rig Package	112
5–9	Triplex Mud Pump Showing the Three Cylinder Blocks	115
5–10	Standpipe Manifold On a Land Rig	116
5–11	Mud Tank With Float Type Level Indicator	118
5–12	Hydrocyclone—Principle of Operation	120
5–13	A Bank of 4" Desilter Hydrocyclones	121
5–14	Crownblock Sheaves at the Top of the Derrick	122
5–15	Hoisting System Schematic	123
5–16	Driller at the Rig Floor Controls	124
5–17	Section Through the Rotary Table	124
5–18	Drill Floor and Rotary Table With the Drillstring Hanging In Slips	125
5–19	Drillpipe Suspended In the Rotary Table Using Slips	127
5–20	5" Drillpipe Elevator, Open	128
5–21	Drill Crew Torquing Up a Drillpipe Connection	130
6–1	Types of Drill Bits	133
6–2	PDC Bit, Roller Cone Bit With Tungsten Carbide Teeth, Natural Diamond Bit and Roller Cone Bit With Steel Teeth (clockwise from top left)	135

Figures

6–3	PDC Cutter Mounted On Tungsten Carbide Stud	136
6–4	Bit Profiles (diamond bits)	137
6–5	Fishtail Bit	138
6–6	Core Sample	139
7–1	Montmorillonite Crystal Composition	148
7–2	Rheology—Shear Rate	155
7–3	Consistency Curve for a Newtonian Fluid	156
7–4	Consistency Curve for a Bingham Plastic Fluid	158
7–5	Consistency Curve for a Pseudoplastic Field	159
7–6	Consistency Curve for a Dilatent Fluid	160
7–7	Consistency Curve for Herschel-Buckley Fluids	160
8–1	Whipstock Used to Deviate a Well	175
8–2	Bent Sub Above Motor	176
8–3	Downhole Motor With Bent Housing	177
8–4	Rebel Tool Conceptual Drawing	180
8–5	Horizontal Well Typical Profile	182
8–6	Multilateral Well Diagram	183
9–1	Buoyancy Effect On Casing	194
9–2	Buckling In Casing or Tubing	195
9–3	Cement Flow In an Eccentric Annulus	198
9–4	Casing Centralizer	199
9–5	Cement Slurries Used Outside Casing	202
9–6	Cement Plug Container	209
9–7	Diagram Showing How the Stinger Seals In the Casing Float Shoe	210
10–1	Explosive Sidewall-coring Bullet	223
10–2	Cores Taken With a Rotary Sidewall-coring Tool	224
10–3	Modular Dynamics Testing Tool for Sampling Fluids and Pressures	225
10–4	Cement Evaluation Tool Log	227
10–5	Borehole Geometry Tool Log	228
10–6	Logging On Drillpipe	231
10–7	Graph of Wellbore Pressure Vs. Time During Drawdown	234
10–8	Pressure and Derivative Curves for Different Reservoirs	235
11–1	Cutaway Diagram of a Bag Preventer	240
11–2	Replacement Rubber Element for a Bag Type Preventer	241
11–3	Set of Fixed Pipe Rams	242
11–4	Variable Bore Ram (one of a set of two) Showing the Sealing Element	242

Drilling Technology in Nontechnical Language

11–5	Blind-shear Ram (one of a set of two) Showing the Flat Seal and the Blade Below	243
11–6	BOP Stack Assembly for a Land Rig (one Bag, two Ram preventers)	244
11–7	Choke Valve—Conceptual Drawing	245
11–8	BOP Control System for a Surface BOP Stack	247
11–9	Kick Situation Diagram	250
11–10	Kill Phase 1 Pumping Pressure	253
11–11	Diverter and Side Outlets	255
11–12	Phase 1 Kill Graph for a High Angle Well	258
11–13	Rotating Control Head—Williams Series 7000	260
12–1	Typical Drilling Operation Organization Chart	266
12–2	Time-Depth for Well Example 1	271
12–3	Cost Elements by Code	276
12–4	Contingency Costs	277
12–5	Cost Estimate Summary	278
13–1	Profile of a Keyseat	292
13–2	Differentially Stuck Pipe	295
13–3	Drill Collars With Machined Spiral Grooves	296
13–4	Jar—Principle of Operation	298
13–5	Overshot Grapple	300
13–6	Taper Tap	301
13–7	Junk Sub—Conceptual Diagram	301
13–8	Washover Milling Tool With Catch Fingers	302
13–9	Homemade Washover Tool	303
13–10	Flat-bottomed Mill—for Milling Junk On Bottom	304
14–1	Example of a Safety Alert	315

Tables

9–1	Slurry Weight Requirements	202

Introduction

This book was written so that people without a technical background might understand some of the complex issues involved in drilling oil wells. The readership I tried to keep in mind while writing were:

- Managers of enterprises (private, public, or government) who have to deal with drilling operations or operators

- People who work in other parts of the petroleum industry who want to know about the work of their drilling colleagues

- Students who might be considering a career in the industry, or careers advisory officers or teachers

- Spouses of people working in drilling who would like to know what their partner does while they are away from home for such long periods of time

- Anyone else who might simply be curious or interested

Such a technically involved industry as drilling cannot be explained meaningfully without going over some of the math, physics, chemistry, and engineering involved. What I have tried to give you is a detailed explanation of the basic principles in these areas before discussing the various techniques we use to drill an oil well.

I recommend that you read the first four chapters in the order that they are presented. Then you can usefully browse through the other chapters as your interest takes you, with enough technical knowledge to understand what is written.

Unlike most books, this one gives you an easy path to the author. You can log on to my web site at **http://www.drillers.com** where there will be a feedback page for readers to ask questions and download a file of questions and answers. I found this to be very useful and interesting (for me and my readers) with my previous book, *Practical Well Planning and Drilling Manual* (Pennwell, 1998).

Most of all I hope you find the book enjoyable to read. I hate to think of all the work that goes into a book like this resulting in something that sits on your bookshelf, looking impressive and gathering dust!

Thank you.

Steve Devereux
CEng MIMM MIMgt

Chapter 1

Drilling Geology

Chapter Overview

This chapter will examine geology as it relates to drilling operations. It is necessary to understand something about the physical and chemical characteristics of rocks in order to understand drilling processes and problems. The chapter also describes the basic principles of hydrostatic pressure exerted by a fluid at depth.

This brief chapter will cover the most important concepts that should be understood for the chapters that follow. See the list of references at the end of the book for further reading.

Igneous, Metamorphic, and Sedimentary Rocks

When the earth was originally formed, the planet consisted of molten rock. As the surface of the planet cooled down the surface rocks solidified. Rocks that were formed by molten rock solidifying are called *igneous* rocks. Basalt and granite are examples of igneous rock.

As the planet continued to cool, water and gases at the surface formed the oceans and the atmosphere. The rotation of the earth, the gravitational pull of the sun (and its heat) and the moon caused movements of the atmosphere (weather) and the oceans (tides and currents). As the temperature varied between day and night and through the seasons, the rocks expanded and contracted. If the temperature dropped below freezing, water in crevices in the rocks would freeze and expand, breaking pieces of rock away. These effects are called *weathering processes* and they aid the erosion process.

Bits of rock, varying in size from tiny grains to huge boulders, could be carried large distances by wind and water. Eventually the forces carrying the rock particles would reduce and the rock fragments would fall to the earth's surface or to the bottom of a water body, forming thick beds of material called sediments. As the water or wind slowed down the largest fragments would be deposited first, while the smaller fragments (being more easily carried) would move further. In this way, the rock fragments could become sorted so that a particular bed of sediment might consist of fragments all of a similar size. Large particles are deposited in relatively high-energy environments (e.g., a fast flowing river) and small particles are deposited in low energy environments (e.g., a lake or swamp).

Through millions of years, these sediments became buried deep within the earth, where they were subjected to high pressures (from the weight of rock above them) and temperatures (the earth gets hotter at increasing depths). The minerals in the sediments changed; dehydrated and changed chemically by pressure and temperature, they bonded together to form rock. This process is called *diagenesis*. Other types of rock were formed by small grains of minerals becoming bonded together by materials forming where the grains touched. This process is called *cementation*. Rocks that are formed by the diagenesis or cementation of sediments are called *sedimentary rocks*. Shale, sandstone, and limestone are examples of sedimentary rocks.

Apart from these "physical" sediments, chemical sediments also occur. Salt beds formed from the evaporation of water from salty lakes can be very thick, causing some real problems not only for drilling but also for seismic surveying (due to the acoustic properties of salt). Biological sediments, such as fossilized coral reefs and coal, are also significant.

Sometimes, an existing rock (igneous or sedimentary) is subjected to extremely high pressure or temperature, sufficient to modify its crystalline structure or change its chemical makeup. These are called *metamorphic rocks*. Marble (metamorphized limestone) and slate (metamorphized shale) are examples of metamorphic rocks.

In most areas of the world, sedimentary rock lies on top of basement rock (igneous and metamorphic). The layer of sedimentary rock can vary in thickness from 0 (eastern Canada) to 50,000 feet. In areas of volcanic activity, igneous and metamorphic rock may be found at or near the surface with thick sedimentary rock underneath.

Plate Tectonics

Below the thin solid crust at the earth's surface, the planet core is molten. This liquid rock may be seen at the surface in active volcanoes. On top of the liquid rock, the crust consists of plates of solid rock, floating on the molten rock underneath rather like rafts floating on water.

The earth's crust is divided into seven huge, major plates (African, Pacific, Indian-Australian, North and South American, Eurasian, Antarctic) and numerous smaller plates (examples include the Arabian and the Cocos). The plates are all moving relative to one another and at the boundaries between plates, earthquakes are likely. Some plates are moving apart from each other (e.g., North Atlantic and Southern Indian Ocean) and new crust is formed as molten rock exposed by the drift apart is exposed and cools down. Some are converging (e.g., at the western rim of the Pacific) where one plate slides underneath the other and volcanoes and mountains may be formed. Others are sliding past one another (e.g., West coast USA at the boundary of the North American and Pacific). The rate of movement varies from 1.3 cm/year to 17.2 cm/year.

In areas close to the edge of a plate (*tectonically active* areas), the rocks can be under greater stress than in other areas. This can make it difficult to drill a hole that remains stable. The sides of an unstable wellbore will tend to collapse into the hole, enlarging the hole and giving serious problems during oil well drilling.

Movements of the plates will lead to rocks moving up or down within the earth's crust. It will also lead to rock beds becoming folded, broken, and turned over. Pressures of fluids contained within the rocks can be drastically changed from the surrounding rock. Stresses within the rock will be different in different directions.

Lithology

Lithology refers to the physical character of the rock. Lithology is a description of the rock and is based on such characteristics as mineral composition, color, grain size, and other textures. Thus a shale could contain some sand (sandy shale) or a rock could be mainly sand with some shale minerals within it (shaley sand). The lithology will effect many drilling decisions when planning and drilling the well. If the wrong decisions are made due to a lack of detailed lithology knowledge, serious problems can result, which will increase the cost of the well and could even prevent the well from reaching it's objectives.

The constituents of a rock give important clues as to how that rock was formed. A very clean sand with no marine fossils or shales within it may have been deposited originally as a sand dune. Evidence of flows, ripples, and cracks give clues about the environment in which the rock was formed.

Shales

Shales consist of layers of clay minerals. Clay minerals are crystal structures of various metal oxides associated with alumina silicates in connection with varying numbers of water molecules. The metal oxides are most commonly those of iron and magnesium but may also be those of sodium, potassium, or other metals. Their presence together in varying ratios results in a wide range of clay mineral types. Clay minerals originate in sedimentary rocks by the physical and chemical breakdown of other minerals originally present. The weathered clays are carried by wind and/or water to an area of eventual deposition. They may then undergo further physical or chemical breakdown. More water may become associated with the clay minerals.

Eventually the clay minerals become buried under other sediments. They will be compacted and water will be driven off. Diagenesis will alter their structure or composition to change the accumulation of clay minerals into a sedimentary rock type. The water that comes out of the shale is less saline (contains less salt) than the water left behind, so that the shale gets more salty as it dehydrates.

Some shales react very quickly with water. Hydration of these shales leads to the crystalline layers expanding. During drilling, fluids are pumped down the hollow drill string and back up the hole to clean and cool the drill bit as well as to perform various other functions. If such a shale is drilled using fresh or salty water as the circulating fluid, the shales will absorb water and rehydrate. Millions of years of diagenesis can be reversed in under an hour. These shales will swell and become soft, like unset putty. Sticky clays will fill the hole and it's possible that the drill string would get stuck and the hole would be lost. Other shale types can be very stable in the presence of water and are not difficult to drill through with water-based drilling fluids (called "drilling muds"). In practice a shale formation may consist of a mixture of different clay mineral types and so the reactivity of shale to water can vary from mild to severe.

Clays are deposited in very low energy environments as they are built from small to very fine particles. In order for these to be deposited, the water must be almost still. Such environments would be found in very deep water, swamps, and lakes.

Shales form about 75% of all sedimentary rocks and cause about 90% of all geology related drilling problems. The successful drilling engineer needs a good understanding of shale chemistry and physical attributes in order to drill wells to economically reach their objectives.

Sandstones

A sandstone consists of particles of sand (mostly quartz grains, often colored by the presence of traces of other minerals such as iron). The sand grains are pressed together by the weight of the sediments deposited above. In the spaces between the grains is water, which may contain all sorts of dissolved minerals and salts. Over time, materials may come out of solution and be deposited where the grains come into contact with each other. This cementation might be strong or weak, depending on the minerals involved and on how much gets deposited. With weak cementation the sandstone may start to fall to pieces when drilled through. If it is a reservoir rock, the flow of hydrocarbons from the formation into the producing well may bring sand with it—which can damage or block production equipment if the well is not designed to prevent it.

Within a sandstone, clay minerals might also be found. Clay minerals cause serious problems in a reservoir if they react with water (perhaps from the drilling fluid) because they can block the flow of hydrocarbons to the well.

Well-cemented sandstones may also be quite abrasive, leading to high rates of wear on the drill bit and other downhole components.

Sandstones make up about 11% of all sedimentary rocks.

Carbonates

These are composed of fossilized skeletons and mineral grains of calcite (crystals of calcium carbonate, chemical formula $CaCO_3$). Crystalline limestone is very common and its texture can be seen by rotating the rock and observing the light reflected off the numerous crystal faces. The crystals are soft enough to be scratched by a knife.

Fossiliferous limestone is similar to crystalline limestone except that it contains fossil fragments usually composed of calcium carbonate. Most limestones are fossiliferous.

Carbonates are often fractured due to their brittle nature. Fractured carbonates make prolific reservoir rocks. Drilling through fractured carbonates can cause large volumes of drilling fluid to be lost into the formation. Sometimes these formations have to be drilled with special techniques such as using foam as a circulating medium rather than liquid mud.

Limestones often contain chunks of *chert* (flint). This is an amorphous quartz, i.e., without crystals. It forms from percolating silica-rich pore fluids. The rock breaks along curved surfaces forming knife-edges and sharp points. Cherts can break teeth on drill bits while drilling.

Carbonates comprise about 13% of sedimentary rocks.

Evaporites (salts)

Evaporite sequences occur as a result of seawater evaporating, leaving the soluble salts behind. There is a definite order of precipitation as the least soluble salts come out of solution first. If a saltwater lake evaporated without further influx of salt water, the order of precipitation would be calcium carbonate ($CaCO_3$), dolomite ($CaMg(CO_3)_2$), gypsum ($CaSO_4 \cdot 2H_2O$)—which is converted to anhydrite ($CaSO_4$) as heat and pressure remove the associated water molecules—halite (NaCl), and lastly, various rare potassium and magnesium salts. These latter salts are very soluble and would only precipitate out if dehydration was almost complete.

The complex sequences present in a mixed salt formation lead to several problems which cannot be solved by using a conventional salt (NaCl) saturated water-based drilling mud as a circulating fluid. The highly soluble magnesium and potassium salts will dissolve in a sodium chloride saturated solution. This can give greatly enlarged holes with attendant problems of lost directional control, difficult cementing, and difficult or impossible fishing operations if a fish[1] gets lodged across a large washout.

Thick salt formations also create some interesting problems. Under high pressures, salt flows just as ice in a glacier flows. It can create tremendous forces that act on any obstacle in it's path, such as a well. It is possible for flowing salt to break a well in half or to crush the steel casing that lines the well, under the right conditions. Salt can flow so fast that a hole can close around the drill bit as it drills, stopping the bit from turning.

Often when that happens, fresh water has to be pumped down around the bit to dissolve some of the salt in order to free the bit.

As salt is lighter than most other rocks, bubbles of the salt can try to rise up through the rock above it, like a bubble of oil rising through water. Of course it takes millions of years for this to happen, rather than a few seconds! These salt domes can be huge and can create traps for hydrocarbons. Salt domes create suitable conditions for many reservoirs in the Gulf of Mexico.

Originally almost all sediments would have been laid down as horizontal layers (or "beds") as shown at the top in Figure 1-1. The "bedding plane" is horizontal. Different types of sediment will be laid down in different environments and as the environment can change over time, so the sediments form a sequence of different lithologies. For an undisturbed sequence of layers, the oldest layers are deposited first, at the base of the sequence, and the youngest last, at the top.

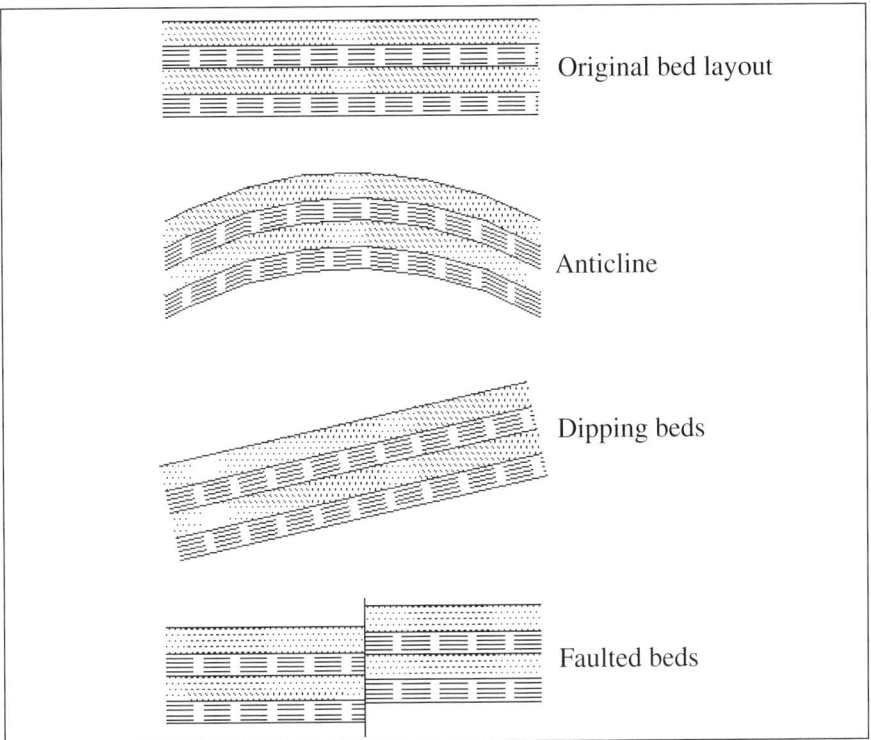

Fig. 1-1 Bedding Formations

As the sediments become buried deeper, forces on the beds (perhaps from tectonic movements or uneven settling) can distort the beds. Distorted sedimentary rock beds can be seen in most parts of the world, at cliff faces, in quarries, and at other places where the rock beds are exposed. One distortion that is of interest is an anticline, where the rock has been folded into a kind of arch. Where the rock is distorted in the opposite direction, like a bowl, the structure is called a syncline. Depending on how brittle the rock is, the time that the distortion takes to occur, and other factors, the rock beds (also called *strata*) might or might not fracture.

Sometimes strata can stay reasonably straight but are tilted so that they dip at some angle to the horizontal. The bedding plane, originally horizontal, now lies at an angle to the horizontal. It is also possible for strata to be turned completely over so that younger sediments lie below older ones!

Where the forces in the rock act in different directions over small distances, the beds may shear so that they no longer line up. This break is called a *fault*. Faults may provide a good, pressure tight seal (in deeper strata), or they may allow fluids to move up along the fault.

Rock Strengths and Stresses

The strength of a rock will vary depending on the type of stress applied to the rock (compressive, tensile, or shear) and may also vary depending on the direction that the stress is applied. Most rock has little tensile strength—it will pull apart relatively easily. However, rock can have great compressive strength, especially if the compressive force is applied at right angles to the bedding plane of a sedimentary rock. Compressive strength is important—a rock with high compressive strength is harder to drill through, but it also tends to be more stable (less likely to fall to pieces) once it has a hole drilled through it (Fig. 1-2).

When a rock is buried in the earth, it is subjected to stresses from the rock around it. The weight of the overlying rock (the overburden) will apply a vertical stress to the rock. In a normal area, where there is little or no tectonic activity, the horizontal compressive stress will tend to be around the same magnitude as the vertical compressive stress and will be of a similar intensity from all directions. However, where tectonic or other forces act to distort the rock, stresses may differ depending on the direction that the stress is measured. Salt domes will impose severe directional stresses on the surrounding rock. These unequal stresses may lead to failure of the rock in

Chapter 1 • Drilling Geology

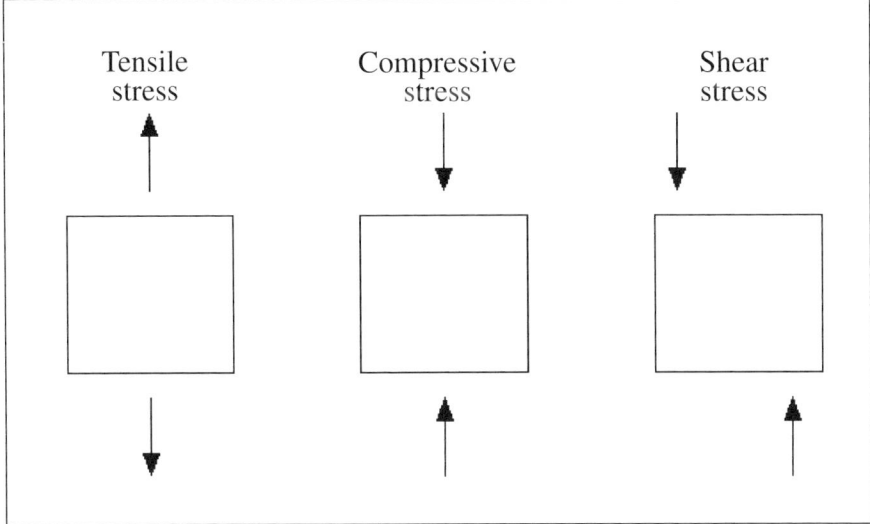

Fig. 1-2 Comparison of Compressive Strengths

the direction of the greatest stress, so that the hole is less stable in one direction—the hole may become oval or eye shaped.

Shear stresses may lead to faults in the rock, as discussed previously. In some cases, drilling into a fault causes some further movement along the fault and this can literally grab the drill bit and stop it from turning—though this is a rare occurrence.

Wells are sometimes drilled straight down vertically, however most wells deviate from vertical to a greater or lesser degree and wells may even be drilled so that they finish up horizontal in the reservoir. In deviated (and especially highly deviated or horizontal) wells, these stresses can become a significant factor in designing the well and deciding on the procedures needed to drill through it successfully can be challenging.

Hydrostatic Pressure Imposed by a Fluid

Fluid pressures are fundamental to many aspects of oil well drilling. If downhole pressures are not kept under control, an uncontrolled release of oil and gas to the surface (called a *blowout*[2]) can result, which might lead to loss of life, massive environmental damage, damage to underground reservoirs, and damage to the rig and other surface facilities.

9

Drilling Technology in Nontechnical Language

Fluids (liquids and gases) exert pressure in all directions, against a vessel containing the fluid and against anything submerged in the fluid. This principle is used in hydraulics, where a fluid is used to transmit a force from one place (a hydraulic pump) to another (a hydraulic jack used to lift a car).

In a vertical column of fluid, gravity causes the pressure inside the fluid to change with depth. The pressure in the fluid at a particular depth has to support the weight of the fluid above that depth. Let's explain this one step at a time.

Imagine a liquid that weighs 0.0417 lbs for each cubic inch of volume. This liquid is stored in a 12" long square tube that has sides of 1" x 1". See Figure 1-3 below.

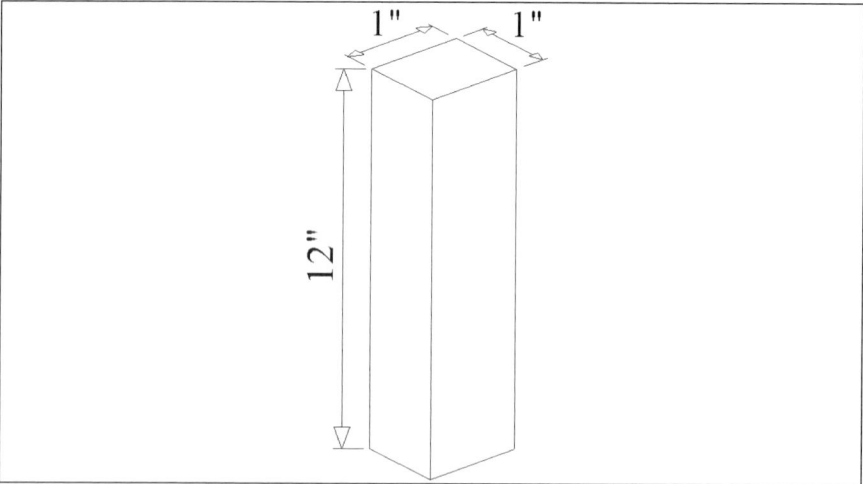

Fig. 1-3 Square Liquid Storage Tube

Now the volume inside this tube is 12 cubic inches (12" x 1" x 1"). If the weight of 1 cubic inch is 0.0417 lbs then the weight of the 12 cubic inches inside the tube is

12 x 0.0417 lbs. = 0.5 lbs. (half a pound)

Pressure is calculated by dividing a load (in pounds) by the area supporting that load (in square inches). In this case, the area at the bottom of the tube is 1 square inch, so the pressure at the bottom of the tube is

0.5 lbs ÷ 1 square inch = 0.5 pounds per square inch (psi)

Halfway down the tube there is half of the weight of fluid (1/4 lb) sitting on the same area (1 square inch) so the pressure is half of what it is at the bottom of the tube. If the depth halves, the pressure halves, and conversely if the depth doubled to 2 feet then the pressure would also double to 1 psi. For each foot of depth, the pressure increases by 1/2 psi. This allows a very easy way to calculate the pressure at any depth—for this fluid, it's simply the depth in feet multiplied by 1/2 psi. The pressure gradient is 0.5 psi per foot (psi/ft).

Fresh water has a pressure gradient of 0.433 psi/ft and salt water (though it varies depending on the amount and types of salt dissolved in the water) is around 0.465 psi/ft. In a well which was 10,000 feet deep and full of salt water with a pressure gradient of 0.465 psi/ft then the pressure on bottom of that well would be 4650 psi. Easy!

In a stack of permeable rocks from the surface to 10,000 feet, if there are no permeability barriers (such as a layer of salt or clean shale) then the pressure at 10,000 feet can be calculated if the pressure gradient of the fluid inside the rock pore spaces is known. In an area where the pressure is "normal" then the fluid in the pore spaces will average about 0.465 psi/ft.

It is possible for pressures in a formation to be much higher than is normal for the depth (this is termed *overpressured*). For this to occur, two conditions are both necessary

1. There is a pressure tight barrier above the overpressured formation
2. There was a mechanism that created the higher pressure

It is not necessary to go into the various mechanisms that might cause overpressures under a barrier, it's only necessary to accept the possibility of overpressures. When such an overpressured formation is drilled into, if the formation pressure is higher than the hydrostatic pressure of the drilling mud, then the mud is pushed up the well by the pressure in the formation. This is called a kick. If this happens then the rig crew must seal the top of the well to stop formation fluids entering the well and then replace the mud in the well with a heavier fluid which gives a hydrostatic pressure greater than the formation pressure. This process is called *killing the well*. A well that is under pressure from a formation is said to be live.

If the well encounters a kick that is not controlled, the well will blow out. Hydrocarbons will flow freely to surface, may ignite, will endanger the rig and the people on it, and will cause huge costs and pollution.

It's very important to know how the pressure in the rock pores might vary with depth in the planned well. In a known area where other wells exist, the pressures and depths will be known, but even so, surprises may still occur. It's vital to never be complacent about downhole pressures or to assume that everything that is likely to happen is known, no matter how many wells are already drilled in the area. Most of all, the situation must never occur where a well isn't strong enough to resist the pressures encountered during a kick.

Chapter Summary

This chapter examined the basic geological processes that affect how wells are drilled. It discussed how rocks are formed and showed the basic rock classification (igneous, sedimentary, and metamorphic). How downhole stresses may vary and how this affects the stability of the wellbore was mentioned. It also summarized the main sedimentary rocks (shales, sandstones, carbonates, and evaporates) and briefly mentioned some of the drilling problems that might arise while drilling through them. Finally, the principles of hydrostatic pressure were explained and what happens when the mud hydrostatic is less than the formation fluid pressure.

Glossary

[1] **Fish.** Something in the hole that should not be there and has to be removed before drilling operations can continue. If the drillstring breaks, then left in the hole, the drillstring and drill bit below the break would be considered a *fish*.

[2] **Blowout.** A situation where formation fluids (oil, water, gas, or a mixture) can freely flow out of the well and into the environment. This situation is extremely dangerous; it is likely to lead to loss of life, loss of the rig, loss of the well, damage to the environment, and damage to the reservoir. The chance of fire is high. The flow might include H_2S, which is fatal at low concentrations if breathed. Stopping a blowout requires specialized skills and equipment. There are several companies worldwide who specialize in blowout control, the most famous being started by Red Adair (now retired).

Chapter 2

Oil and Gas Generation, Migration, and Reservoirs

Chapter Overview

Hydrocarbons are rarely found in the rocks where they were generated from the organic remains of plants and animals. As hydrocarbons are generated, they migrate upwards within the earth until they meet a barrier, where they may accumulate as an oil or gas reservoir. If they do not meet a barrier they will eventually reach the surface where they will evaporate, possibly leaving behind the heaviest hydrocarbon elements (e.g., the tar sands of Alberta, Canada).

There are seven recognized factors that all need to be present, in the correct order, for a petroleum accumulation to occur. These are known as the *Seven Pillars of Wisdom for Oil Accumulation* and are in this sequence:

1. Source rock
2. Hydrocarbon generation
3. Primary migration to a suitable structure (reservoir)
4. Structural trap
5. Reservoir rock
6. Seal rock
7. Secondary migration within the reservoir

Drilling Technology in Nontechnical Language

This chapter will briefly describe the processes involved in petroleum generation, migration, and accumulation into an exploitable reservoir. The important rock properties for reservoir creation and the properties of the fluids within the reservoir are described.

Source Rock and Hydrocarbon Generation

Millions of years ago, plant and animal remains could sometimes accumulate and become buried by deposited material. Often this was in slow moving or stationary water environments such as swamps, lakes, coastal regions, and shallow seas. As these organic remains and other non-organic particles (clay minerals, fine sands, and silts) sank to the bottom, thick beds of sediment built up over a long period of time (thousands to millions of years).

As the organic-laden sediments became buried deeper by more deposited sediments and by movement of the earth's crust, they were subjected to increasing temperatures and pressures. The sediments, under diagenesis, became sedimentary rocks. About 99% of all hydrocarbon deposits are found in sedimentary rocks. Within the tiny pore spaces of the rock, the organic matter also underwent chemical transformation.

Under certain conditions, oil is generated from organic remains. The most important factor is temperature; oil generation starts at 50°C, conversion to oil peaks at 90°C and stops at 175°C. This range of temperature, 50° to 175°, is known as the oil window. As higher temperatures generate oil more quickly, younger sediments must have been exposed to higher temperatures to create significant quantities of oil. However, even at higher temperatures, oil generation still takes millions of years.

Below and above the oil window, decay of the organic remains will generate gas. Below 50°C, biogenic gas (generated by microbes) or swamp gas will result. Above 175°, thermal gas will result. Temperatures at the lower end of the oil window will generate heavy oils, with increasing temperatures generating lighter (and more valuable) hydrocarbons. If the temperature of the rock becomes too high (above 260°C) then the organic material (and therefore the oil generating potential) is destroyed.

Petroleum comprises carbon (83%) and hydrogen (13%), and sometimes small amounts of sulfur (2%), nitrogen (0.5%), and oxygen (0.5%). Hydrocarbons (composed of carbon and hydrogen only) make up over 90% of most crude oils. The hydrocarbons in crude oils vary in molecular size and molecular type, with the heavy crudes comprising more large molecules and the light crudes comprising smaller and more volatile molecules.

Rocks that produce hydrocarbons from organic matter buried within the rock pore spaces are known as source rocks. The most common organic-rich sedimentary rock, thought to be the source rock for most oil and gas, is shale. Many shales are black and are often referred to as black shale. The black color comes primarily from its organic content.

A black shale may have 1%-3% of organic matter, whereas a green or grey shale may have only about 0.5% organic matter. Organic-rich black shale is relatively common in many areas of the world (e.g., the "Kimmeridge Clay" shale formation of the North Sea, through which the channel tunnel linking Britain and France runs). Of the organic matter deposited in the earth's crust, only about 2% becomes petroleum. Of this 2%, only about 0.5% (that is, one part in 10,000 or 0.01% of the original organic matter) finds its way into a commercially exploitable reservoir. Thus the petroleum deposits we want to exploit come from only a tiny fraction of all the organic matter that is deposited within the earth's crust. Petroleum generation is a very inefficient process.

Coal comes from the deposits of woody plant remains. Peat and poor-quality "brown" coal has been subjected to lower temperatures and pressures, while high quality black coal (e.g., anthracite) has been buried deeper and exposed to higher temperatures. As the woody remains convert to coal, hydrocarbon gas (methane) is generated, which can migrate upward to form a gas-only reservoir (no associated oil). This methane can also remain trapped within the coal seams, where it presents a serious danger to the miners extracting the coal.

Vital Rock Properties

Petroleum is generated within small voids, or pore spaces, of a source rock. Many rocks have pore spaces within them. Sandstone comprises grains of predominantly quartz, chemically cemented together with minerals that accumulate where the grains touch (Fig. 2-1).

The mineral cement holding these grains together may be very strong and the rock is then described as being *highly* or *well consolidated*. However, in some cases this bonding is not very strong and such a rock would be described as *poorly consolidated* or *unconsolidated*.

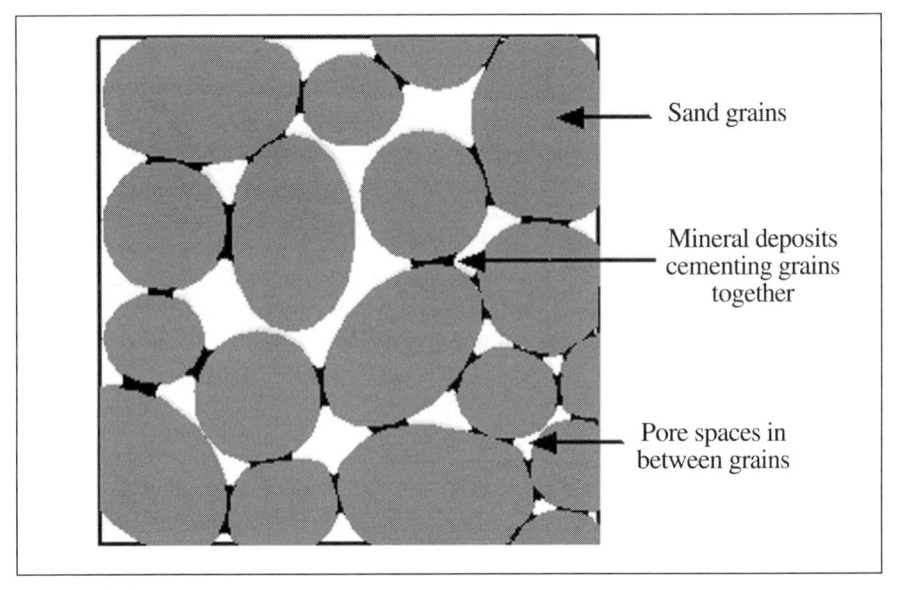

Drilling Technology in Nontechnical Language

Fig. 2-1 Typical Sandstone Structure

Materials that have spaces within them, like a sponge, are described as *porous*. The extent of this porosity is measured as the fraction of the total rock volume occupied by the pore spaces. Porosity is expressed as a percentage; if 20% of the total volume of a rock was made up of pore spaces, the porosity would be 20%. Rock porosity is very important, for without porosity, oil cannot be generated, cannot migrate, and cannot accumulate in a reservoir.

Within a porous rock, it is possible for the pore spaces to be connected. Fluids (gas, oil, or water) can flow between the pores, moving through the rock. However, there are also rocks that are porous where the rock spaces are not connected. Fluid in the pore spaces cannot move through the rock. The ability of a rock to allow fluid to flow through it is called *permeability*. A rock can be porous but impermeable (cement is an example of a porous, impermeable solid), but a permeable rock must have porosity (pore spaces that can be connected must exist). Permeability is extremely important, for without it oil generated in the source rock cannot migrate to a reservoir and cannot be exploited.

Permeability is measured in *darcies*. A rock cube with 1cm x 1cm x 1cm sides that transmits fluid with a viscosity of one centipoise at a rate of

1cc per second with a pressure differential of 1 bar has a permeability of one darcy. In layman's terms, a rock with a permeability of one darcy is very permeable. Most reservoir rocks are measured in millidarcies (1/1000th of a darcy) rather than darcies.

Shale has very low porosity and very low permeability. Shale minerals form flat crystals that stack up like plates on a shelf. When clays are originally deposited, they are comprised of 70%-80% water. As water is squeezed out of the clay during diagenesis, these flat crystals become very close to each other. At a depth of 2000m, the gap between the crystals is about 10 nanometers. By the time the shale is buried to a depth of 5000m, this gap has closed to about 1.5 nanometers (1.5 x 10^{-9} meters or 0.0000000015 meters).

Shale has tiny pores that are connected by tiny passages. It takes a long time for the water and the oil produced within the shale source rock to be squeezed out by pressure. The actual mechanism by which the oil leaves the source rock is uncertain, but it is thought that the oil is initially in solution in the water under the high pressures that exist in the source rock.

Primary Migration

The first two necessities for the birth of a reservoir are the existence of an organic rich source rock and the conditions necessary for oil to be generated; temperature (the oil window) and time. If the oil cannot migrate out of the source rock, it stays locked within the shale and cannot be produced.

The third element required is for the source rock to lie next to a permeable rock or a channel that allows the oil to migrate. In most cases, a permeable sandstone deposit provides this conduit, but it can also be provided by fractures in the rock or ancient reefs (limestone structures made up of coral skeletons with very high permeability). Fractures often allow migration vertically upwards and this mechanism has led to many large oil accumulations, such as those found at shallow depths in Venezuela and Northern Iraq.

If a sandstone formation allows the oil to migrate, a gently sloping formation bed can carry the oil for long distances horizontally until a trap stops migration and allows accumulation. Therefore, a reservoir can lie many miles away from the source rock that generated the oil.

Structural Trap

As the oil and gas undergoes primary migration away from the source rock, it must find a structure that has the right conditions to trap the oil and stop it from reaching the surface.

A trap may be formed by movements within the earth's crust that deform the rock (anticline). This gives a kind of inverted cup structure. Thick salt deposits can flow under high pressures and sometimes a bubble of salt rises from the salt bed to form a salt dome. Salt domes distort the rocks above them and can provide large structural traps suitable for a reservoir. Reservoirs may also be formed by permeable beds that are tilted by crust movements, eroded after exposure to the surface and buried by new layers of impermeable rock deposits. This is called an *angular unconformity* (an unconformity is a break in the natural rock layer). Figure 2-2 illustrates these structures.

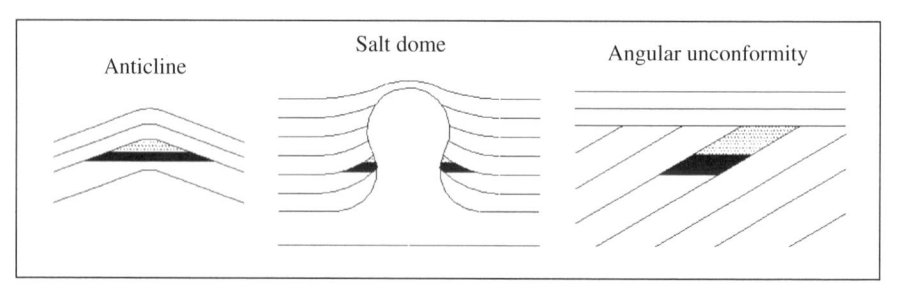

Fig. 2-2 Stratigraphic Traps

Reservoir Rock

In a reservoir, the gas, oil, and water are found in pore spaces or fractures within the rock matrix. Most reservoirs worldwide are contained in sandstone structures that have sufficient porosity to give a good volume of reserves and a high enough permeability to be able to produce it. However, limestones (carbonates) with fractures and/or pits (due to water dissolving some of the material) can give extremely high porosity and permeability. Some limestone reservoirs originate from coral reefs and tend to be prolific reservoirs, with high porosity and permeability. In the U.S., around 80% of reserves are held in sandstone and 20% in carbonate reservoirs. In the Middle East, almost 100% of reserves are in carbonates.

Reservoir permeability ranges from a couple of hundred millidarcies to 15 darcies or more. The more permeable a reservoir, the faster the hydrocarbons within it can be produced.

Reservoir rocks often contain other materials within the pore spaces. For instance, in a sandstone reservoir there may be varying amounts of clay minerals. These can cause problems while drilling through the reservoir. The clays can react with part of the drilling fluid and expand to plug pores and passages, reducing permeability in the zone around the wellbore. This can seriously reduce or even prevent production of oil and gas from the well.

Reservoirs are rarely uniform throughout. They may consist of layers of material with slightly different characteristics, leading to "directional permeability"—the permeability is different depending on the direction of flow. Permeability might be better horizontally than vertically. Within the reservoir there may be faults or distortions. Other rock types may be present. These things all form barriers to the free flow of hydrocarbons and they can make the reservoir structure extremely complex. In fractured limestones, the fractures containing oil may be vertical, requiring a wellbore to be drilled horizontally in order to intersect many fractures for efficient production. Clearly, selection of where to place a wellbore in order to hit the best (most permeable) parts of the reservoir can get complicated!

Modern 3D seismic techniques can obtain detailed knowledge of the reservoir structure and are now indispensable tools in modern well planning. This data and data from other wells drilled in the reservoir during exploration are used to create computer simulations that model reservoir structure and behavior. This in turn allows the operator to exploit the reservoir with the minimum number of wells and surface facilities—maximizing return on the money invested in developing the field.

Seal Rock

While a reservoir rock must be permeable, there must also be an impermeable rock seal above it that prevents further upward migration of the oil and gas. Often this seal is formed by a layer of "clean" shale (shale with little or no sand within it). Other impermeable seal rocks may be formed from salt or unfractured limestone.

In chapter 1, the possibility of drilling into a formation that had a greater pressure than would be expected for the depth of the formation was described. A seal rock gives one of the two conditions necessary for overpressure to occur—a pressure barrier.

Secondary Migration

In secondary migration, the oil droplets move about within the reservoir to form pools. Secondary migration can include a second step during which crustal movements of the earth shift the position of the pool within the reservoir rock.

Accumulations can be affected by several (sometimes conflicting) factors:

1. Buoyancy causes oil to seek the highest permeable part of the reservoir; capillary forces[1] direct the oil to the coarsest grained area first, then successively into finer grained areas later

2. Any impermeable barriers in the reservoir can channel the oil into a somewhat random distribution

3. Oil accumulations in carbonate reservoir rocks are often erratic because part of the original void spaces have been plugged by minerals introduced from water solutions after the rock was formed

4. In large sand bodies, barriers formed by thin layers of dense shale may hold the oil at various levels; with crustal movement of the earth, accumulations are shifted away from where they were originally placed

5. Faults (where part of the rock body has moved along a crack) sometimes cut through reservoirs, destroying parts or shifting them to different depths

6. Uplift and erosion bring accumulations nearer to surface, where lighter hydrocarbons may evaporate

7. Fracturing of the cap rock may allow accumulations to migrate upwards

Wherever differential pressures exist and permeable openings provide a path, petroleum will move.

Some (very few) reservoirs are *single phase*—i.e., they contain only a single fluid type, either all gas or all oil. Oil is rarely found without some gas or water. Gas generated from coal seams, as described previously, can be single phase. However, reservoirs are mostly *multi-phase*—they contain mixtures of gas, oil, and water. Secondary migration will tend to separate these fluids out by gravity so that the gas sits at the top (known as a gas *cap*), oil rests under the gas, and water lays under the oil (lightest fluids at the top to heaviest fluids at the bottom). The oily part of the reservoir may contain a mixture of oil and water within the pore spaces, in which case the reservoir rock may have a layer of water adhering to the surface of the rock grains (water wet) or it may have a layer of oil adhering to the rock (oil wet). These factors and others must be considered by the reservoir engineers when deciding how to exploit the reservoir.

Reservoir Drives

What provides the energy to drive the hydrocarbons to the surface through a hole drilled into the reservoir? Most oil reservoirs, when first drilled into and produced, have sufficient pressure in the reservoir to push the oil to the surface. This energy can come from different sources.

Gas drive

The reservoir is partially or completely isolated from the pressure regime in the surrounding rock. As oil is produced the gas cap above it expands. As it expands it loses energy—the temperature and pressure in the reservoir drop until eventually there is not enough energy left to drive the oil out. Gas drive is not an efficient long term production driver.

At the stage when there is insufficient pressure left, there are several possibilities

1. Inject more gas into the reservoir to increase the pressure
2. Ignite oil underground by injecting air; the gas from the burning oil increases reservoir temperature and pressure and drives more oil out

3. Install a downhole pump to pump the oil to the surface; this may be mechanical (with a "nodding donkey" or "horsehead" on the surface providing the power via rods connected to the pump), a downhole hydraulic, or a downhole electric pump
4. Inject gas into the well; this mixes with the oil, which makes the column of oil lighter, allowing the reduced pressure to drive the oil out (this is called *gas lift*)
5. Inject water and chemicals in part of the reservoir to drive the oil towards the producing wells

These techniques are called *secondary recovery*.

Water drive

In the previous chapter, the principle of hydrostatic pressure was explained. Liquids have a pressure gradient—the heavier the fluid, the higher the pressure gradient. In a sedimentary rock sequence there is a local "water table", i.e., the rock pore spaces contain salt water and this exerts a pressure at depth (explained in chapter 1).

In a reservoir that has water drive, the reservoir is connected hydraulically to the area pressure regime (such as an aquifer that is open to the atmosphere). Water from the local water table pushes the oil to the oil well. Water drive maintains the driving pressure much longer than gas drive as the pressure comes from the surrounding area. Eventually, as the oil is produced, water will move towards the well and at a certain point it may break through the oil and reach the well. The water will then be produced in preference to the oil because it is less viscous and because the increase in the amount of water in the pore spaces blocks the oil—this is called *water blocking*.

Problems Related to Fluids in the Reservoir

In some cases, the oil in a reservoir can become degraded by bacterial action to produce hydrogen sulfide (chemical symbol H_2S). H_2S is extremely toxic; exposure to more than 100 parts per million of H_2S in air can be fatal. While that is bad enough, steel exposed to H_2S in a wet envi-

Chapter 2 • Oil and Gas Generation, Migration, and Reservoirs

ronment can become brittle and fail (break) without warning. If a reservoir has H_2S present, the rig must be equipped to deal with it. Crews must be trained and equipped to work safely in the presence of H_2S. Expensive steel alloys must be used in the well to prevent equipment failure. H_2S is also called *sour gas*. It has a distinctive odor of rotten eggs, but H_2S quickly destroys the sense of smell.

Carbon dioxide (CO_2) can also be present in the reservoir. Exposure of steel to CO_2 in a wet environment can also cause serious corrosion problems for steel tubes in the well.

If both H_2S and CO_2 are present in a reservoir, there are serious problems in production as well as treatment of the produced oil or gas before it can be sold.

Chapter Summary

1. For an exploitable petroleum reservoir to form, specific conditions arising in the correct sequence are necessary.

2. Economically exploitable hydrocarbon reserves are generated from about 0.01% of all organic matter deposited within the earth's crust.

3. Reservoirs may be formed from permeable sandstones or permeable/fractured carbonates. The oil and gas is found in the pore spaces, fractures, and voids within the reservoir rock.

4. Reservoirs can be extremely complex, requiring detailed structural knowledge and computer simulations to work out how best to exploit them for the maximum return on investment.

5. When planning and drilling wells, close cooperation between the drillers and the reservoir engineers is necessary in order to minimize damage to the reservoir.

6. If H_2S is present in the reservoir, special precautions are needed to avoid danger to the crews and the well has to use special, expensive steels to avoid catastrophic sudden failure.

Glossary

[1] **Capillary** forces arise due to the attraction between fluids and solids. In a very small passage (capillary), these attractive forces can cause the fluids to move along the passageway—even vertically upwards against gravity. Capillary forces are responsible for "rising damp" in houses when the bricks are porous and have some permeability.

Chapter 3

Planning and Drilling an Exploration Well on Land

Chapter Overview

This chapter will tell how a land rig exploration well in the desert might be planned and executed. An exploration well is one that is drilled in order to prove whether or not a structure in the rock contains hydrocarbons in commercially worthwhile quantities.

Some technical terms (not too many, I hope) will be introduced and explained in this chapter. In addition, many of the concepts that were covered in chapters 1 and 2 will be referred to.

Identifying a Prospect

An operator buys the right to explore for oil or gas in an area from the government. These areas are often called *blocks*. The operator[1] will buy a permit to explore on a block because they believe that:

1. All of the conditions discussed in chapter 2 are likely to exist for hydrocarbons to accumulate in commercial quantities

2. The likely cost of exploration is worth the chance of finding an exploitable field

3. They can afford to absorb the exploration cost if nothing worthwhile is found, or can afford to develop any discovery in order to bring hydrocarbons to market

Drilling Technology in Nontechnical Language

It may be that previous work was done on the block and hydrocarbons were encountered. Or perhaps the geology of the area gives cause for optimism. In the early days of the oil industry, "wildcat" exploration wells[2] were drilled simply because somebody had a "gut feeling", or if there was oil or gas seeping from the earth. In modern exploration, the easily found shallow fields have probably all been discovered.

An exploration well is drilled to gain information. It is usually false economy to try to drill an exploration well in order to later produce oil. A producing well cannot be properly designed until the reservoir is known in sufficient detail (pressures, fluids and gases present, permeability, how well consolidated the reservoir rock is, and many other factors). Many things about the subsurface conditions cannot be predicted on the first well. This means that the well design may have to change if unexpected conditions are encountered while drilling. Exploration wells should be minimum cost wells designed to obtain essential information and then to be abandoned[3].

Well Proposal

For our example well, an angular unconformity structure is present (as was discussed in chapter 1), providing a potential stratigraphic trap. The seismic survey indicates that oil, gas, and water might be present with a gas cap at the top. It was decided to drill a well into the edge of the gas cap and down through the oil into the water, so that several facts could be established (the well objectives):

1. Prove that oil and gas both existed in the reservoir, obtain samples of each for analysis, and measure the fluid pressures in the reservoir
2. Determine the depth of the gas-oil and oil-water contacts[4]
3. Take core samples[5] in the oil part of the reservoir
4. Test the oil layer to measure the following:
   a. the maximum rate at which the oil can flow before sand starts to be produced
   b. the maximum possible production rate
   c. internal reservoir characteristics[6] such as permeability, porosity, internal boundaries, pressures, and temperatures

Chapter 3 • Planning and Drilling an Exploration Well on Land

    d. damage to the reservoir from the drilling operation—the mechanical skin[7]

Now a *well proposal* document has to be written. This is a request from the exploration department for a well to be drilled and it provides the necessary information to the drilling department for the design of the well.

Well proposal contents

1. Type of well (exploration) and well objectives (as stated above)
2. Essential well design data:
   a. surface location, if known
   b. downhole targets to hit—position, depth, and the acceptable margin for error
   c. depths and descriptions of the downhole rock strata (as far as can be determined)
   d. expected strengths of formations downhole and pressures of fluids inside the rocks
   e. what information is required from the well
   f. whether core samples of rock should be obtained and at what depths
   g. what should happen to the well after operations are complete (abandoned, or perhaps temporarily secured for possible later use)
   h. outline of the completion design for the well (see below for an explanation)

First, the drilling department needs to review the well proposal to establish that

1. The proposal is logical and the objectives are achievable
2. The essential well design data is complete and there are no ambiguities or omissions
3. The directional targets that the well needs to hit are as large as possible; the smaller the target, the more the well is likely to cost

4. The proposal doesn't give rise to any inherent hazards that might create a danger to personnel, the rig, or to the well

Once all concerned parties agree upon the well proposal, the drilling department will look for any available information to help the drilling engineers design the well.

Completion design

Once a well is drilled, steel pipe is lowered into the hole, right through the reservoir, and cemented in place. Holes are blown through the casing, cement, and into the formation with special explosive charges (*perforations*) to allow oil or gas to flow into the well.

In order to flow hydrocarbons to the surface, tubing is run into the well. This connects to a special tool, called a *packer*, which seals the space between the casing and the tubing, preventing hydrocarbons from flowing outside of the tubing (Fig. 3-1).

The tools that are installed inside the casing to flow the well, including the completion tubing, packer, and any other tools, are collectively known as the *completion*. The completion design, and especially the required size (diameter) of the completion tubing, must be known before the well can be designed because the diameter of the completion tubing dictates the final hole and casing diameters.

The drilling engineers are not responsible for designing the completion. This is the work of either the production or exploration department. If it is an exploration well, the completion is then designed for testing the well, which is then known as the *test string*.

Gathering Data

The well design defines the final status of the well—what will be physically present once the drilling rig has finished its work and left the wellsite. In order to create an efficient design for the well, as much as possible must be known about the conditions above and below the surface.

In locations where other wells (*offset wells*) have been drilled reasonably close to the new well, information can be obtained about likely formation characteristics and how to deal with them.

Sometimes, when formations are drilled, a side force is exerted on the bit. The formation characteristics are such that the bit is pushed in a par-

Chapter 3 • Planning and Drilling an Exploration Well on Land

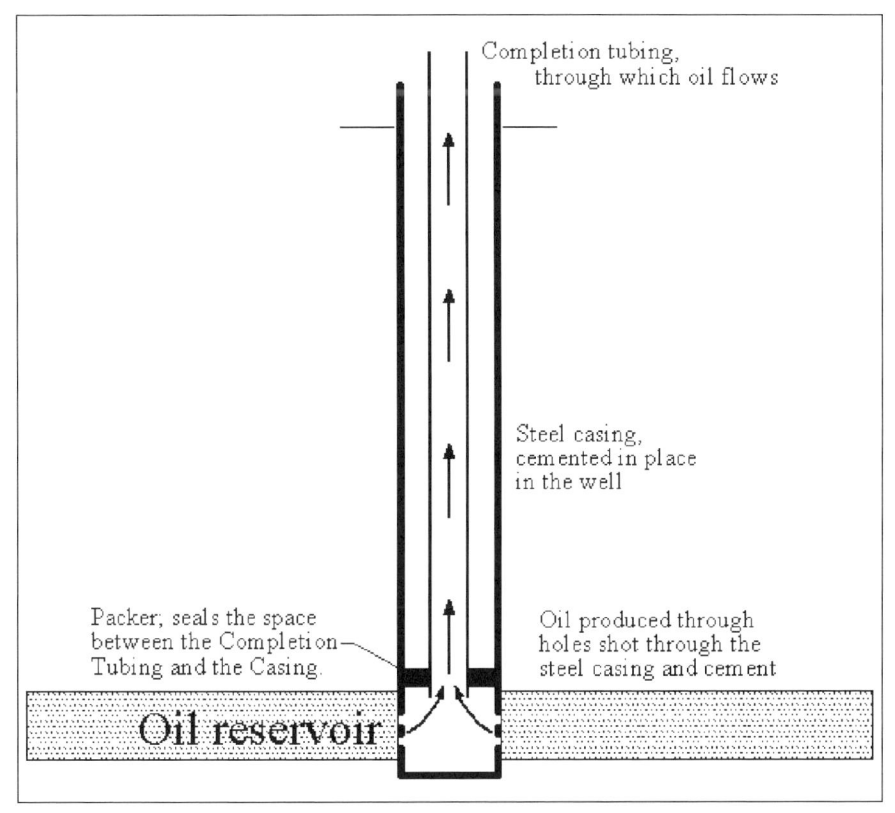

Fig. 3-1 Illustration of a Packer

ticular direction as it drills. Imagine trying to drill a hole in a plate of steel, but the drill is not square to the surface of the steel. The drill would tend to skip along the surface of the steel, unless it could be held firmly in place with something like a jig.

Imagine a bit drilling through some dipping beds like the ones shown below (Fig. 3-2). The beds alternate in hardness. As the bit hits each hard formation, the bit will try to move off to the left until it has drilled deep enough into the formation to cut a full circle. Over a distance of hundreds or thousands of feet, this will make the bit wander off course. If in a 7,000 feet deep hole the bit only deviates by 10° in the top 1,000 feet and then drills straight, the target would be missed by almost 1,000 feet!

Each formation sequence drilled will have its own particular directional characteristics. If these can be determined, the well can be designed to hit the target by following these natural tendencies as much as possible. In the case of a vertical well, there will still be characteristics of the rocks that may tend to cause deviations from the vertical.

All of this subsurface data has to be gathered, collated, summarized, and presented in ways that are useful while working on the well design. Computerized databases and other software tools are very useful, but so are other methods of working with the data. It doesn't *all* have to be high tech.

So much for the subsurface part of the well design. What other information will be useful? As this chapter concerns a mythical exploration well drilled in the desert, it might be expected that the well can be drilled from any surface location. A simple vertical well would be ideal. However, the terrain above the target might not allow a rig to be placed there. Access roads may have to be built. If a local water supply is not available, a water well has to be drilled; otherwise, all water has to be brought in by truck.

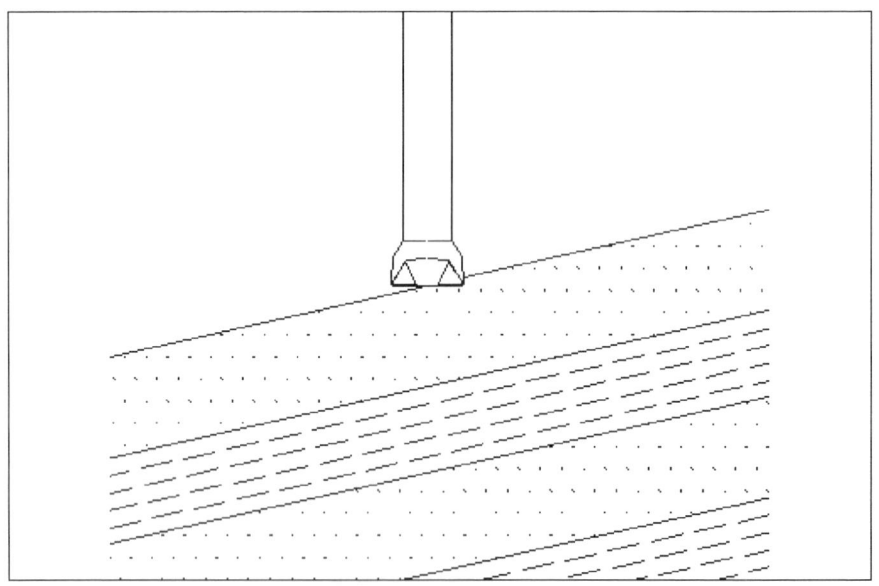

Fig. 3-2 Bit Drilling through Dipping Beds

The environmental impact of the operation must be considered in advance. Often, authorities now demand an "Environmental Impact Assessment" of the proposed site, which is a detailed survey of the flora and fauna around the site. Drilling a well generates waste materials (solids, liquids, and gas emissions) and there is always a risk—however slight—of an unplanned discharge of hydrocarbons. Noise, lights, and heavy traffic may disturb local wildlife behavior patterns. It may be necessary to take special precautions in sensitive areas, such as removing all solid and liquid waste for disposal rather than burying it on site. These requirements will increase the cost of the well, which has to be estimated at the start in order to get budget approval to drill the well.

In many Third World countries, apart from the political authorities, local tribal leaders have to be consulted in order to gain land access and to operate without interference. This is likely to involve payment of a "compensation", but may also involve employing local people as laborers.

Designing the Well

Hole sizes

The well design "defines the final status of the well". The first thing to determine is the diameter of the hole through the formation. This drives much of the design and also has a huge influence on the final cost. For an exploration well, the final hole size will usually be the larger of what is required for a) running electrical measuring tools (*called logs*) or b) running tubing into the well to allow hydrocarbons to flow for a well test.

Figure 3-3 demonstrates how a well progresses in a series of "hole sections" which are drilled in progressively smaller hole sizes. Casings are run to consolidate the current progress, to protect some zones from contamination as the well progresses (such as freshwater sources) and to give the well the ability to hold higher pressures. Note that the top of each casing string is set higher than the top of the previous casing string, so the well grows in height as more casings are run.

While there will be an ideal minimum hole size to obtain all of the desired information, it is possible that the *essential* information could be obtained in a smaller size hole. A well might be planned to finish in 8-1/2" diameter hole (which would allow all of the logs and tests to be run) but it might also be possible to run the bare essential tools in a 6" diameter (or even smaller) hole. This allows for a contingency—if there are unanticipated problems while drilling and an extra set of casing pipe has to be set, the essential objectives can still be met, though some of the "nice to have" information may be sacrificed.

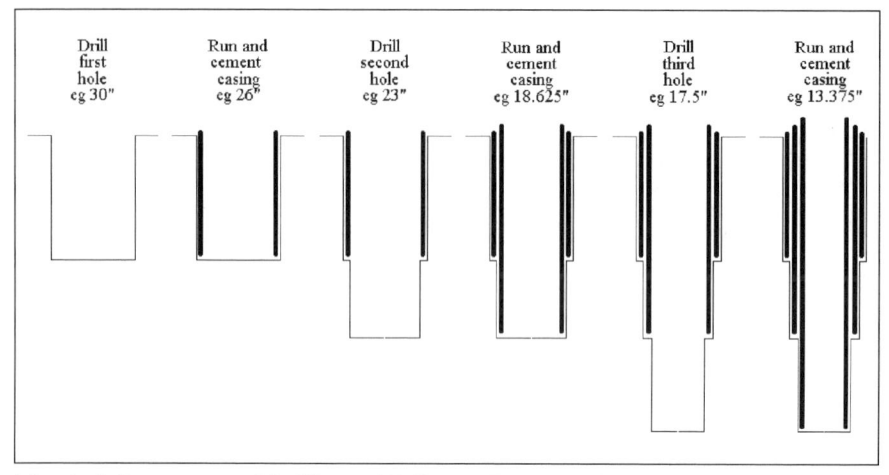

Drilling Technology in Nontechnical Language

Fig. 3-3 Sequence of Drilling a Well and Running Casings

In an exploration well, many things are unknown. Compared to a development well (drilled in a known area), there is a greater risk that problems will occur and an extra casing string might be needed to isolate a problem formation. In an exploration well, a contingency hole size is important to allow the vital objectives to be reached even if this happens. The alternative is to risk losing the complete well (not reaching the vital objectives) if serious problems are encountered.

Casing design

The bottom of a string of casing is called the *casing shoe*. The formation in which the casing shoe is cemented is very important because it must be able to withstand pressure in the event of a kick (as explained in chapter 1). If a kick happens and the well is closed at the surface using the blowout preventers, extra pressure is exerted all along the well from the kicking formation. A simple example illustrates this.

> A well is drilled to 5,000 feet and casing is set. Drilling continues to 10,000 feet. The mud has a gradient of 0.5 psi/ft so the hydrostatic pressure at 10,000 feet is 5,000 psi. A formation with a pore fluid pressure of 5,500 psi is penetrated so the hydrostatic pressure from the drilling fluid is 500 psi less than the formation pressure. With the BOP closed, 500 psi shows on the pressure gauges. This *extra*

Chapter 3 • Planning and Drilling an Exploration Well on Land

pressure will then be exerted everywhere on the well. By reading the surface gauges and knowing the well depth and drilling fluid gradient, the pressure inside the kicking formation can be easily calculated:

Pore pressure in the kicking formation =
surface pressure + hydrostatic pressure

At the casing shoe depth of 5,000 feet, the mud hydrostatic pressure is 5,000 x 0.5 = 2,500 psi. If a kick occurs and the surface gauge reads 500 psi, then the pressure on the casing shoe will also have increased by 500 psi—now it will be 3,000 psi and not 2,500 psi. So it's vital that the formation at this casing shoe can withstand at least 3,000 psi. If it breaks, there is a very serious problem. The pressure might break formation all the way to surface allowing pressurized hydrocarbons to travel up, out of the seabed or the ground around the rig, where it might ignite.

The formation that the casing shoe is cemented in should, ideally, be strong and impermeable—clean, unfractured shale is perfect. Limestone may also make a good casing shoe formation, but limestone is usually brittle and can fail with no warning when pressure is applied to it. Salt can be adequate if it's not flowing too much and if it's not a complex salt sequence prone to serious hole enlargement. Well-consolidated sandstone, which has low permeability, may also be used. The casing shoe should not be cemented in a formation that is fractured, unconsolidated, very permeable, or of low strength.

By examining the sequence of formations that should occur while drilling the well, those that meet the criteria for setting the casing shoe can be marked. In a genuine wildcat exploration well, this will not be known before drilling and it may then be a case of drilling until a suitable formation is drilled into, then setting the casing before drilling further.

There is a maximum depth that can be safely drilled below any particular casing. This depth depends on

1. The formation strength at the casing shoe
2. The density of the drilling fluid in the well
3. The hole diameter

4. The maximum volume of formation fluid that can be allowed into the well (called the *influx volume*)
5. The density of the formation fluid that enters the well in a kick

It's not necessary to go into the calculations for this, but it is important to know about it. These calculations are called *kick tolerance* calculations because they show how large of a kick—a combination of formation fluid pressure and influx volume—a well can withstand without breaking down the formation at the casing shoe.

By knowing which formations are suitable for the casing shoe and calculating how far to drill before the kick tolerance becomes too small, the ideal depth to set each casing can be determined. This also dictates how many different hole sections (and casings) are needed.

Designing the casing strings is quite a complex job. Currently, computer programs are used to arrive at the most cost-effective solution that meets the requirements of the well.

Writing the Well Program

Once all of the relevant data is collected and a well design to meet the well objectives is agreed upon, the next job is to write a program. While the well design shows the final status of the well, a well program will instruct the rig on how to implement that design. Think of the well design as being like a set of blueprints for an aircraft, but in order to build that aircraft the factory workers need to have a wide variety of instructions, procedures, and the design drawings.

The well program contains, among other things, a set of advisory instructions to the rig which show how it is thought the well can be drilled most efficiently. Note especially the word "advisory". It is impossible to anticipate what will happen as the well is drilled. The drilling supervisors in charge of the rig operations may need to deviate significantly from the program if safety or efficiency might otherwise suffer. A well program should never be thought of as a precise set of instructions, but rather as advice that, if justified, can be changed. However, it is also important that the program contains the information behind all of the major decisions made, so that all of this original information can be combined with new information to make the most informed decisions possible. In other words, the program should not just say what should happen, but also *why* it should.

Chapter 3 • *Planning and Drilling an Exploration Well on Land*

A well program is a comprehensive document and can be quite long. The major headings of a well program for drilling and completing a well may include the following sections:

1. General information
2. Well objectives
3. Potential hazards
4. Surface location and how the rig is to be positioned
5. General notes, including:
    a. references to government regulations, company policies, oilfield standards
    b. reporting procedures
    c. quality control and data recording requirements
    d. a diagram of the completed well
    e. equipment checklists and suppliers of each item
    f. cost estimating information to allow the well cost to be calculated each day
6. Drilling notes for each hole section including:
    a. potential hazards or problems, how to avoid them, and how to recover from them
    b. required drilling practices
    c. recommended operational sequence of events
    d. kick tolerance information
    e. what drill bits are recommended
    f. what bottom hole assemblies are recommended (this is explained below)
    g. any special requirements
7. Drilling fluid design and maintenance requirements for the whole well
8. Wellbore trajectory information (the path that the well will follow from the surface location to the downhole target[s])
9. Casing design for the well and how the casings are to be cemented in place
10. Geological information on the formations expected to be penetrated

11. Logging and coring program—what electrical logs to run and, if the well is to have a core sample cut, the relevant details (Logging and coring are discussed in chapter 10)
12. Well completion design and program
13. Well test information, if the well is to be production tested
14. Status of the well when the rig has finished work (e.g., handed over to production, cemented up and abandoned, etc.)

As you can see, there is a lot of information contained in a drilling program! Each heading is examined briefly below.

General information

1. Which country, which exploration block (blocks are usually numbered), name of drilling rig, program issue date, who the program was written by, and who approved it
2. Which offset wells were used for data input
3. A statement on shallow gas[8] (e.g., whether it's likely to be present or not)

Shallow gas

Sometimes a pocket of gas can be found very close to the surface (or the seabed). On many wells, the first string of casing (called a *conductor pipe*) consists of very thick walled pipe, often more than an inch thick. This is usually hammered into the ground with a pile driver. The purpose of the conductor is to "conduct" drilling fluid back up to the rig when drilling first starts—normally water with an additive to make it viscous. The first hole section—called *surface hole*—is drilled and afterwards the "surface casing" is cemented in place.

Until the surface casing is in place, the well does not have enough strength to shut in on a kick. If a deposit of shallow gas is drilled, the well cannot be closed in and circulated to a heavier fluid in the conventional manner. The well is allowed to flow through a special type of blowout preventer called a *diverter*. This merely diverts the flow away from the rig and its purpose is to buy time for the crews to get off the rig. With luck, the gas deposit is small and soon depressurizes, or rocks produced with the gas (and there will be a lot!) will jam in the hole and stop the flow. Otherwise the chances of destroying the rig are high.

Figure 3-4 is a diagram of the shallow gas situation, showing the shallow gas deposit that might be hit before setting surface casing into the ground.

Shallow gas presents the most serious type of danger to the rig and the people on it. If shallow gas is hit and the well kicks, things will happen very quickly. Within a few seconds, all of the drilling fluid in the well can be blown out so that gas is flowing freely out of the well. Many rocks or large volumes of sand or other material will come out with the gas and this can erode steel lines very quickly, allowing gas to flow within the rig structure. Shallow gas may also contain hydrogen sulfide.

To avoid hitting shallow gas, some precautions can be taken:

- Carry out a special type of shallow seismic survey[9], which should indicate the presence of shallow gas
- If the well is offshore, first drill a small hole at the proposed well location from a special kind of floating rig that will quickly move away if shallow gas is hit
- Examine the records of offset wells and see if shallow gas was encountered in any of them

Well objectives

It's important to differentiate between primary objectives (those which the well must meet) and secondary objectives (those which are desired if they can be obtained for little extra effort/cost).

A graph is normally given, showing the anticipated well depth at each day of the operation. The actual progress can be plotted on the same graph, to show whether the well is on target, "behind the curve" (late) or "ahead of the curve" (early). The flat spots on the graph show where drilling stops to run casing into the well at the end of each hole section (Fig. 3-5).

There will also be a detailed cost estimate for the well, broken down into various categories so the actual costs can be later compared to the estimated costs for each category.

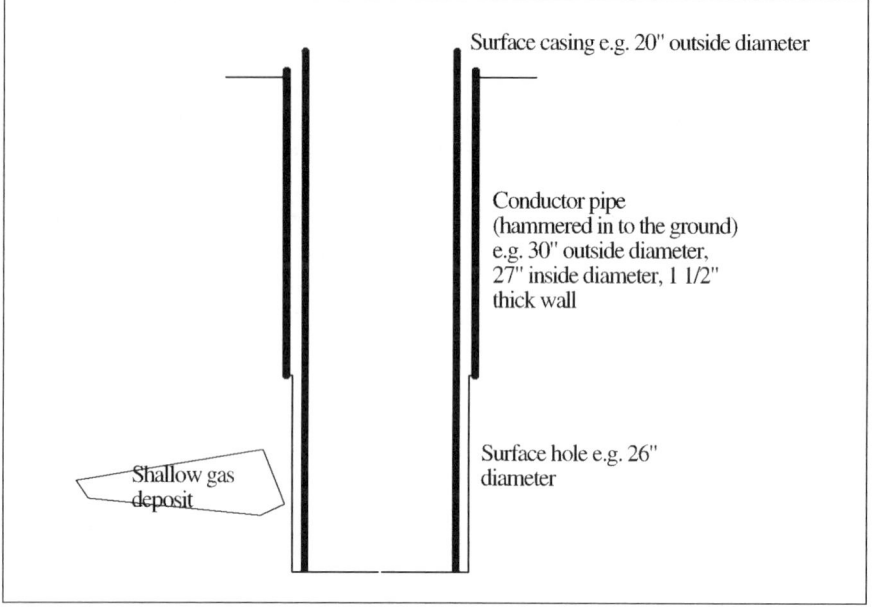

Fig. 3-4 Diagram of Shallow Gas

Potential hazards

A list should be given of any potential hazards that the rig might encounter. These are not drilling hazards (which will be listed in the text for each hole section) but hazards that might be inherent to the location. For instance, in some areas, rigs might expect to encounter extreme weather conditions at certain times of the year (such as tornadoes). In such a case, procedures can be given for monitoring potential storms and, under specified conditions, the rig could be evacuated as a precaution after making the well secure.

Surface location and how the rig is to be positioned

A location on the earth's surface may be given by many different reference systems. Latitude and longitude is one coordinate system. There may also be local grids or coordinate systems that the rig has to be positioned to. In any case, whichever coordinate system is to be used must be specified. Also the tolerance (the acceptable error) must be stated. For instance,

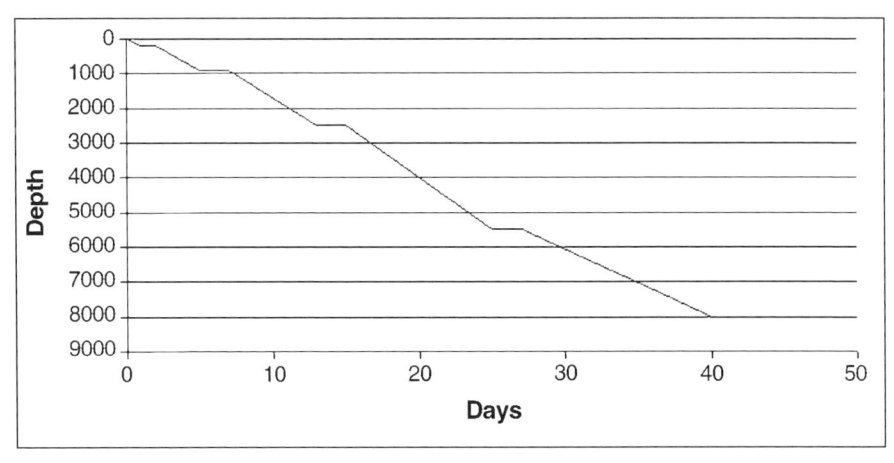

Fig. 3-5 Time-Depth for Well Example 1

"North coordinate 112995.56 East coordinate 122473.88 using X reference system, well position to be within 5m of specified."

The Datum level for depths must be given. This could be Mean Sea Level (MSL), Lowest Astronomical Tide (LAT), or another suitable reference. The water depth for an offshore well must be stated.

General notes

References can be made to relevant government regulations, company policies, or oilfield standards, as appropriate.

A diagram of the completed well showing all relevant design details is usually included.

Drilling notes for each hole section

In each hole section, there may be particular problems that might be expected. A good program would anticipate these problems and offer strategies to avoid or minimize the impact of the problem. It should also offer strategies to recover from the hazard or problem if it were to occur.

Required drilling practices. It may be that specific operational procedures must be used. These can be detailed here.

Recommended operational sequence of events. This gives an anticipated chronological sequence of events, such as drill to a certain depth, run (specified) electrical logs, and run casing.

- What drill bits are recommended?
- What bottom hole assemblies are recommended? (this is explained below)
- Any special requirements?

Drilling fluid design and maintenance requirements

The circulating fluid has a huge impact on almost all aspects of drilling, logging, cementing, and cost. The drilling fluid (mud) must be carefully designed to keep the wellbore stable (a stable hole stays at the same diameter of the bit that drilled it) and to perform many other functions, too. The mud for each section will have specific chemical and physical characteristics that must be kept as designed.

Wellbore trajectory information

Everything that defines the path that the wellbore should follow must be known. Diagrams showing the well profile from the side (called the *vertical section*) and from above (called the *plan view*) must be included. As the well is drilled, the actual wellbore position is marked on the same chart so a glance at the chart shows how closely the actual well path follows the planned well path.

Casing design

Each casing string is completely specified. The outside diameter of the pipe, how thick it is, what sort of threads are needed to screw the casing together, what type of steel will be used to make the casing, and other information. Tools might be screwed into the casing string at particular depths and this must be specified so that the drilling supervisor knows what to put where.

Geological information on the formations expected to be penetrated

In a wildcat exploration well, this information might be scarce or non-existent. In any case there will always be some intelligent guesswork involved, especially when the geologists try to estimate the strengths of the rocks (formation, fracture, gradient) and pore pressure. These two pieces of information are very important to drilling engineers because these are what fundamentally dictate the casing program.

The formation lithologies, permeability, porosity, composition of pore fluids, depths, thicknesses, and the stresses present in the rock are all valuable information.

Logging program

Tools can be run into the well on special electrical cables that measure various formation parameters. These are called *electric logs*. There are several classes of electric logging tools that can be used to identify important formation characteristics and to identify the fluids present in the pore spaces.

Many different formation characteristics can be measured. By combining several measurements, a comprehensive picture of any formation can be built up, identifying physical and chemical characteristics.

Coring program

In an exploration well, it is very likely that interesting formations (those with reservoir potential) are cored. In this process, a special drill bit is used which cuts a doughnut shape, leaving a column of formation sticking up the middle. Behind the bit is a barrel ("core barrel") which contains this core and recovers it to surface.

Logging and coring are discussed in detail in chapter 10.

Well completion design

In order to design the well, the drilling engineers need to know how the well will be completed and, in particular, the size (outside diameter) of the completion or test tubing string. As this is an exploration well, it will use a test string, the size of which can be specified at this time.

Well test information

If the well discovers hydrocarbons, a test program will be written. Although the final test program cannot be known until the hole is logged, it is known in general what the program will involve and what testing equipment will be needed. The equipment has to be arranged in advance and made available to send to the rig as soon as a well test is confirmed.

Status of the well when the rig has finished work

In the case of this example exploration well on land, the well will be abandoned after testing. In the case of other types of wells (such as a producing development well), the drilling engineers need to know what should be the status of the well when they finish their work. It could be left unperforated, or perforated and then killed with clean completion fluid, etc.

Drilling the Well

The drilling program is an important document containing several of the following kinds of information and instructions:

- Compulsory—instructions that must be complied with. Typical instructions will involve safety and reporting procedures. Any deviation from these has to have the permission of the drilling office.
- Advisory—typically things like what drill bits might be best to use at different stages in the well, or mud properties. These things might change as the well is drilled and information gained, or problems encountered that have to be solved.
- General information—such things as emergency contact details that might be needed during the operation.

The general sequence of drilling a well is similar for almost all wells drilled. Events on our mythical well, complete with some typical problems that might be encountered on such a well, are shown below.

Location preparation and conductor driving

On a land location, a preparation crew will build a location in advance of the rig arriving. The site must be cleared of obstructions (and in some areas, such as the western desert in Egypt, these obstructions are unexploded World War II ordinance and mines), the boundary marked and access roads created. A hole will be dug into the ground and lined with concrete. Around this concreted square pit (called the *cellar*) will be a concrete or wooden pad for the rig to sit on.

Why is it necessary to dig a hole under the rig location? On top of the casing is placed a large piece of equipment called a blowout preventer (BOP). Without a cellar for the BOP the rig substructure would have to be much higher to accommodate the BOP.

A large waste pit is dug next to where the tanks of drilling fluid will be placed. Drilled rock cuttings are removed from the well by circulating drilling fluid down the hollow drillpipe, through the drill bit and back up the outside. This drilling fluid lifts up the rock cuttings to the surface. Once at the surface, special equipment (called *solids-control equipment*) separates the drilling fluid from the drilled cuttings. The cuttings have to be disposed of and normally this will be into the waste pit.

With the cellar dug and cement lining in place, the location preparation crew will use a crane and a pile driver to drive the conductor pipe into the ground, in the middle of the cellar. A flange is welded on top of the conductor so that the diverter[10] can be attached before drilling starts.

A pile driven conductor is generally driven until it can't be driven any further. This is called *driving to refusal*.

In a remote desert location, a concrete lined water pit is dug and a water well drilled. Water is required in large quantities during drilling, to mix drilling mud and cement as well as to keep the rig clean. The water well produces water into the water pit. If the water well cannot keep up with the rig requirements, water can be trucked in and dumped in the water pit to supplement the water well production.

Ordering equipment

A lot of equipment is needed to drill even a shallow well on land. Depending on the country, equipment may be freely available (Europe or the U.S.) or it may take months to order, ship, clear customs, move to the logistics base, inspect, and be ready to mobilize the rig. In some countries, equipment can be tied up in customs for weeks or months. It also happens in some places that customs officials expect a bribe to be paid before equipment is cleared or else further delays are incurred.

If an operator or contractor is established and well experienced in the country, then these factors will be known and can be allowed for. With long lead time[11] items, it may be necessary to get special approval to order those items before the well itself is approved.

Quality control is very important, to ensure that items ordered are as per specification and are fit for purpose. If equipment fails (breaks or stops working) while drilling, it can cost a lot of money and in some cases can endanger personnel working around it. There are various industry standards, ISO 9000 requirements, and company procedures that can assist the drilling engineer in ensuring that equipment ordered is suitable.

Checking the infrastructure

Infrastructure covers roads, airfields, harbors, storage areas, etc. A drilling operation requires a lot of logistical support and the infrastructure should be able to handle high traffic loads, at any time of the day or night, in all weathers. If there is an emergency situation and if the infrastructure cannot support the demands of the rig in solving that emergency, then people can be placed in more danger, more environmental damage may occur, and perhaps the reservoir can be damaged if the situation is not resolved quickly.

Moving the rig on location and attaching the diverter

Land rigs break down into packages that are moved by truck. To reassemble the rig on location requires care and precision. Each major part must be accurately positioned relative to the rig substructure, so that cables, walkways, pipes, etc. line up and can be easily connected. Each rig has a procedure outlined in the rig Operating Manual that details "what must go where" and in what order for the most efficient assembling of the rig.

The rig substructure is constructed from steel beams welded together. The substructure is a large frame which supports the drill floor, generally about 15-20 feet above ground level, and the derrick (also less commonly called the *mast*). The substructure itself may split into several smaller

Chapter 3 • Planning and Drilling an Exploration Well on Land

packages for transport. In the middle of the drill floor is a hole, inside of which is a powered turntable. This is called the *rotary table* and is used to turn the drillstring for drilling, as well as for other tasks.

Once the substructure is placed in the correct location, the derrick is laid out onto cradles and assembled. The cradles position the mast so that it can be attached to the substructure above the drill floor. At the substructure end, the derrick is secured by large steel pins, several inches in diameter, which allow the derrick to pivot to the upright position. When everything is ready, the derrick is rotated to vertical by winching in on steel cable attached to the derrick and the remaining two legs are pinned in place onto the rig substructure. With the derrick upright, rigging up continues until the rig is ready to start drilling.

The drillers measure the distance from the drill floor to the top of the conductor pipe. Depths in the well while drilling are referenced back to the drill floor, so the conductor shoe depth equals the length of conductor in the ground plus the distance from the drill floor to the top of the conductor (Fig. 3-6).

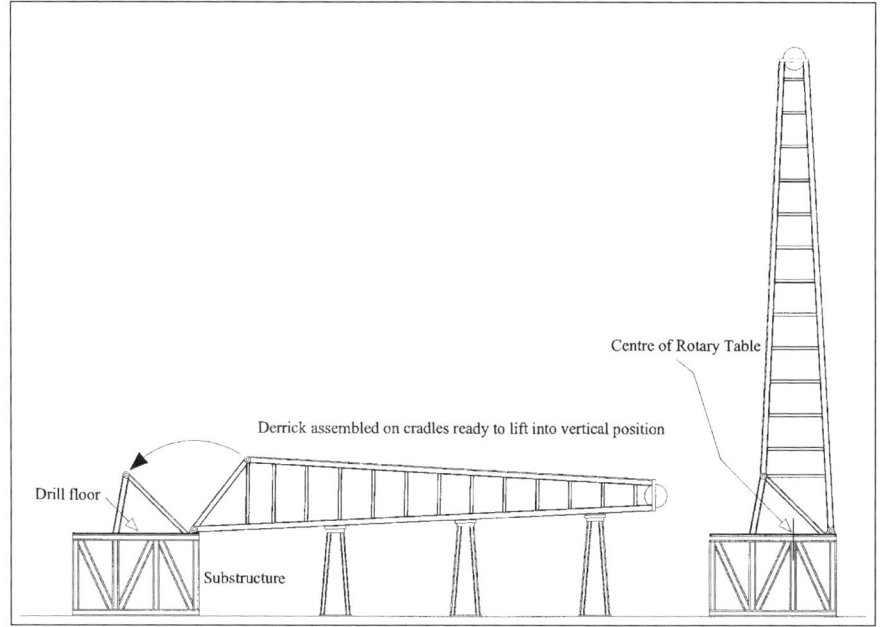

Fig. 3-6 Diagram of a Land Rig Derrick Being Erected

Drilling Technology in Nontechnical Language

The derrick has several large sheaves at the top end. Steel wire rope, called *block line*, passes over these sheaves and around another set of sheaves on a massive pulley. By winching in or out on the block line with an electrically powered drum, the pulley—called the *travelling block*—moves up and down the derrick. Below the travelling block is a large steel hook that can lift whole strings of casing pipe, support the drillstring while drilling, and perform many other tasks. A large land rig would probably be strong enough for the travelling block to support up to 500 tons, using block line of commonly 1-5/8" diameter with a tensile strength of over 100 tons (though block line may vary in size from 1" to 1-3/4" diameter).

With the rig ready to start operating, the diverter must be attached to the conductor that was pile driven in position by the location preparation crew. The diverter contains a large rubber seal that is forced under hydraulic pressure to squeeze in around the drillstring and seal around it. Underneath this seal are usually two large pipes, at least 10" in diameter, which should lead away from the rig in opposite directions with no bends or changes in internal size. Occasionally only one line will be fitted, leading off downwind of the prevailing wind. If a kick is experienced while drilling below the conductor pipe, the flow is diverted away from the rig by closing the diverter and opening the valve on the pipe leading down wind (Fig. 3-7).

On top of the diverter is a section of pipe (called a *bell nipple*) which has an outlet to the side. This side outlet directs mudflow from the rig along a channel to the solids-control equipment and then back to the mud tanks, where the pumps circulate it back down the hole.

Spudding the well[13]

A conductor pipe that was driven into the ground with a pile driver will be full of rubble, as the bottom of the conductor is open while hammering it in to the ground. So the first thing to do is to clean out this rubble. A drill bit that is slightly smaller than the conductor. ID is run on drill collars down to the top of the rubble mud is pumped down the drillstring, and the bit is rotated and lowered. This breaks the rubble up and the mud lifts it up to the surface.

If a shallow gas pocket is penetrated, the chances of recognizing and controlling it are much better if a smaller size hole is drilled. Also, if the well does blow out, it will flow through a smaller hole and the initial flow will be much less. This allows a few vital moments to get the diverter closed and to start pumping fluid down the drillstring as fast as possible to try to slow

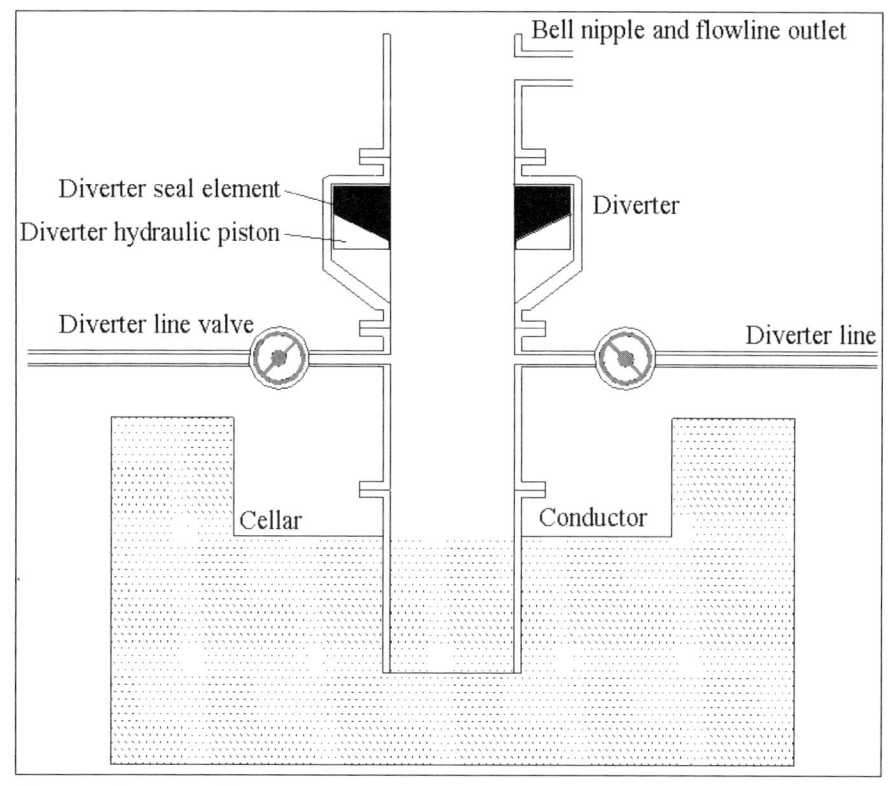

Chapter 3 • Planning and Drilling an Exploration Well on Land

Fig. 3-7 Drawing Showing How the Diverter is Set Up on the Conductor Before Drilling Starts

or stop the flow. The intensity of a shallow gas blowout is such that large volumes of rock will be removed from the well as the gas flows, rapidly making the hole much larger.

The first drill bit will be 12-1/4" diameter. The hole it drills is termed a *pilot hole* because the hole will be re-drilled later with a larger bit. The start of drilling the well is called *spudding* the well. The time that the well is spudded, by convention, is taken to be the time that the drill bit exits the bottom of the conductor pipe.

To drill rock, a drill bit requires a downward force to be applied, to force the teeth to penetrate into the rock. This force is provided by the weight of thick walled pipe, called *drill collars*, which are screwed on top of the drill bit. With a large drill bit used to spud the well (perhaps a 23" or

Drilling Technology in Nontechnical Language

26" diameter bit inside 30" conductor pipe), these drill collars might be 9-1/2" outside diameter and 3" inside diameter steel pipe. Each foot of this particular drill collar will weigh approximately 200 lbs and drill collars come in lengths of approximately 30 feet—so a single drill collar will weigh approximately 6,000 lbs. As the bit drills deeper, more drill collars are added on top, giving more available weight to drill with.

A drill bit also requires to be turned, in addition to having weight applied. With sufficient "weight on bit" and "rotary speed", the bit will drill rock. Stronger rock requires greater weight on bit to be applied, so that the pressure exerted by each tooth is greater than the compressive strength of the rock. Close to the surface, rock is usually fairly soft and easily drilled so having low weight available is not a problem.

As the bit drills, rock cuttings are generated. These must be removed from the wellbore. This is achieved by pumping fluid down the hollow drillstring, out of holes in the bottom of the drill bit and back up the space between the hole and the drillpipe. The space between the drillstring and the hole wall is called the *annulus*. As the drilling fluid or mud flows back up the annulus, the rock cuttings are carried up also and are separated at the surface by the solids-control equipment (Fig. 3-8).

This fluid will most often be water with various chemicals mixed in, but it may also be an oil-based fluid, a mixture of oil and water, or in some areas where conditions permit, it might be compressed air, or a mixture of water and air (which produces a foam). The fluid used to drill the top section of the well is called *spud mud* because it's the drilling mud used in spudding the well. Spud mud is usually water which has clays and polymers added to make it thicker (more viscous).

The speed at which the fluid moves up the well is measured in feet per minute (or meters per minute on a metric operation). This speed measurement is called the *"annular velocity"* (usually abbreviated "AV"). To lift cuttings upward, depending on how thick or viscous the mud is and depending on its density, a minimum AV of about 50 feet per minute (FPM) is needed. More than 100 feet per minute is preferred to efficiently clean the hole. The number of gallons each minute needed to give any particular AV can be easily calculated. The larger the hole size, the more gallons each minute must be pumped.

For a hole of 26" diameter (D) and using 5" diameter drillpipe (d), the AV equals $0.0408 (D^2 - d^2)$ gallons per foot, or 26.6 gallons per foot. To achieve an AV of 50 FPM, the mud flow has to fill 50 feet of hole each

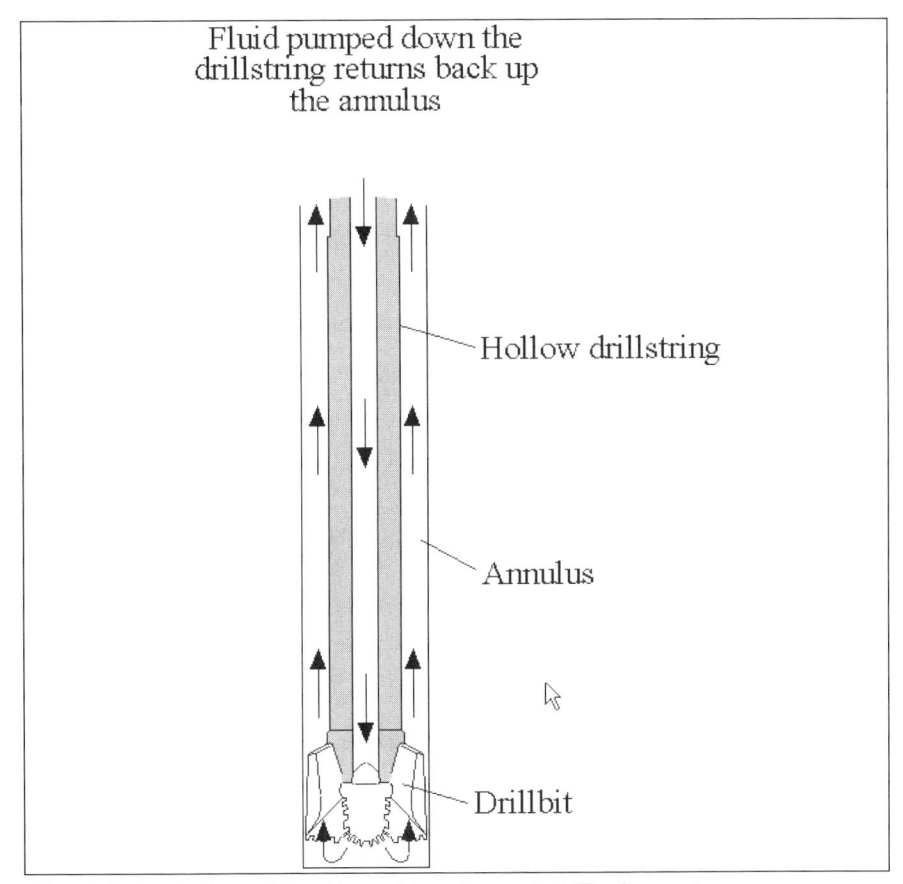

Chapter 3 • Planning and Drilling an Exploration Well on Land

Fig. 3-8 Illustration of the Use of Solids-control Equipment

minute. The capacity of 50 feet of this annulus will be 50 x 26.6 = 1330 gallons. In this large hole, at least 1330 gallons each minute (gallons per minute, usually abbreviated to "GPM") is required to give the minimum AV of 50 (FPM). In our pilot hole of 12-1/4" the flow rate (GPM) is much less; for 50 FPM needs only 255 gallons to easily achieve the preferred 100 FPM. In soft rock, to a limit, the more GPM pumped, the faster drilling can progress, so a 12-1/4" diameter hole is generally drilled with a flow rate of around 600 GPM.

What limits how many GPM can be pumped? Several things. One limit is how fast the pumps can operate at the pressure required. Another is

pumping very fast in softer rock, the force of the mud hitting the bottom of the hole can cause erosion of the side of the hole, making it bigger than the drill bit diameter. A hole that is bigger than the bit (called *overgauge* hole) is undesirable. Also, to force mud up the annulus takes pressure, and this pressure has to be resisted by the wellbore. Circulating faster imposes more pressure on the well. It is possible that a weaker rock could become fractured due to the extra pressure imposed by circulating fast, leading to loss of drilling fluid into the rock.

Drilling the first hole section

After spudding the well, things happen very quickly. The pumps are turning at full speed, all engines are running, and there is a lot of noise. Large amounts of rock cuttings will come out of the well and often the capacity of the solids-control equipment to separate out the solids is exceeded. What happens is that along with the solids being ejected into the waste pit, a lot of mud will also go over the side. Some of the equipment might become plugged up with drilled solids, requiring fast action to get it fixed and back into service. If drilling young clays that react quickly with water, the clay might become sticky, the cuttings start to accumulate into larger lumps, and large balls of sticky clay start to get pushed up the annulus. It can appear at the top of the bell nipple as a column of doughnut-shaped clay wrapped around the drillstring. These are called *clay rings*. Sometimes drilling must stop to clear up the problems before continuing. With clay rings, several people must get down under the drill floor with shovels to clear it. Luckily, modern chemical knowledge is such that clay rings can largely be avoided by proper formulation of the drilling fluid.

After each 30 feet is drilled, another drill collar is added by screwing it on to the top of the drill collars in the hole. When there are enough drill collars to give all of the required weight on bit, drillpipe is added to the drillstring, again in 30 foot lengths. (Each length of drillpipe is called a *joint*.) Before drillpipe can be screwed onto the drill collars, a special short length of pipe with a drillpipe connection on the top and a drill collar connection on the bottom is added. These special short pipes are called subs and if it is used to convert one size or type of connector to another, it is called a *crossover sub*. When the sub connects the drill bit to the lowest drill collar, it is called a *bit sub*. Figure 3-9 illustrates this.

Chapter 3 • Planning and Drilling an Exploration Well on Land

Drillpipe
5" OD x 4.276" ID
x 31' long

Crossover sub to
connect the drillpipe
to the drillcollar

Drillcollar
9.5" OD x 2.875" ID
x 30 ft long.

Drillcollar
9.5" OD x 2.875" ID
x 30 ft long.

Bit sub to
connect the drillbit
to the drillcollar

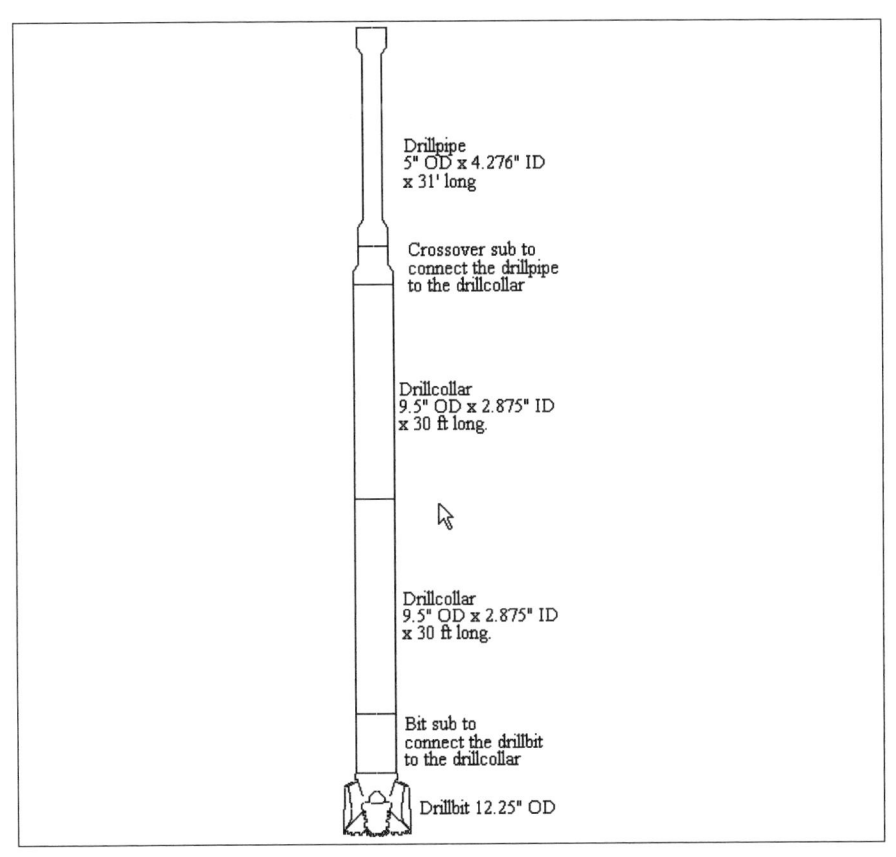

Drillbit 12.25" OD

Fig. 3-9 A Bit Sub

Notice that although the drill collars have straight sides all the way, the drillpipe has a bulge at each end. The drillpipe itself is fairly thin and there is not enough metal to machine a connection onto the pipe itself, so a thick section, with the threaded connection on, is welded to each end of a length of pipe. The pipe part of the drillpipe is called the *pipe body* and the connection part is called the *tool joint*.

Every component added to the drillstring is measured in feet and 1/100th of a foot so it's easy to add all of the lengths together. The record of pipe lengths is called the *tally*.

The components from the drill bit to the bottom of the drillpipe are called the bottomhole assembly (BHA). The BHA can be configured in

Drilling Technology in Nontechnical Language

many different ways to give a particular weight on the bit and to effect how it drills (Fig. 3-10).

Drillpipe tally for well XYZ		
BHA lenth =		296.75
Length	Total	Total with BHA
31.32	31.32	328.07
30.98	62.30	359.05
31.44	93.74	390.49
31.62	125.36	422.11
31.69	157.05	453.80
31.50	188.55	485.30
31.45	220.00	516.75
30.96	250.96	547.71
31.66	282.62	579.37
31.57	314.19	610.94

Fig. 3-10 Example of a Drilling Tally Sheet

The well is drilled until it reaches the depth to run the surface casing.

One problem commonly experienced while drilling a surface hole is mud disappearing downhole into the rock. This problem is called *lost circulation* or, more simply, *losses*. Losses might be slight or severe; losing only a few barrels[14] of mud an hour is called *seepage losses*. Anything more than 30 bbls/hr is moderate and more than 60 bbls/hr is serious. Sometimes the mud losses might be so severe that no mud returns[15] are seen at the surface while pumping over a thousand gallons a minute down the drillstring!

Losses are discussed in more detail in chapter 13.

Once at the required depth, drilling is stopped. Mud is still circulated around until all of the drilled cuttings are cleaned out of the well. If this isn't done and the drillstring is just pulled out, several major problems become likely:

- As the drillstring is pulled out of the hole, the cuttings left in the annulus would fall down and would be very likely to get jammed between the hole and the drillstring. This would stick the drillstring and would also prevent mud from circulating

Chapter 3 • *Planning and Drilling an Exploration Well on Land*

(because the annulus would be blocked). In this condition the drillstring will be stuck and might not be freed, in which case the well will be abandoned.

- Cuttings will settle on the bottom and would make the well shallower than drilled. This might prevent the surface casing being run to the required depth.

- Cuttings might also settle at some intermediate depth (i.e., not on the bottom). When the casing reaches this depth, it hits the debris and starts to push it down ahead of the casing. This would most likely plug the annulus, might plug the casing also and would cause the casing to become stuck.

Whenever drilling is stopped to pull out of the hole, the well must be circulated clean first.

When drilling a surface hole, the formations drilled are generally weak and unconsolidated. The hole might not be very stable—the sides of the hole will tend to crumble slowly (or perhaps quickly) and the hole will enlarge as material falls into the wellbore. This means that the actual size—or volume—of the hole is not accurately known. Material falling off the side of the hole is called "caving" and the bits of rock are called *cavings*. Cavings can be recognized at the surface due to their shape, size and appearance. Careful examination of cavings can give indications of what is causing the wellbore instability.

With the well cleaned up and before pulling out of the hole, a tool is dropped down the inside of the drillstring. This is in the form of a long, slender barrel containing a measuring instrument. The barrel lands[16] on a special landing ring (called a *totco ring*) that is inserted inside the BHA. A clockwork timer mechanism reaches a preset time after dropping (enough for it to reach the totco ring) and the instrument then takes a measurement of the inclination of the wellbore from vertical. In this way, the drillers can tell whether the well is vertical or if it has started to wander off course while drilling. This process is called "taking a survey".

As pipe is pulled out of the hole, only every third drillpipe connection is unscrewed to leave three joints of drillpipe screwed together. The rig's derrick is high enough to stand 90 feet of pipe on the drill floor, secured in special racks at the top. Three joints of pipe screwed together is called a *stand* of pipe. So if there are 45 joints of drillpipe in the hole, only

Drilling Technology in Nontechnical Language

15 connections have to be unscrewed and 15 stands of pipe racked in the derrick. It's much faster that way. The stands are racked in rows. When one row is complete another is started. The drill collars can also be racked in stands of three joints.

After drilling a 12-1/4" pilot hole, it is enlarged to 26" to be able to run in the 20" casing. In a vertical well in softer formations this might be done simply by drilling again with a 26" drill bit. The potential problem with using a drill bit is that if it hits an obstruction (a boulder buried in softer formation) it might be deflected and start to drill another hole. This can be avoided by running a *hole opener*, which has a set of cutters mounted outside a short drill collar. Below the hole opener is placed a tool called a *bullnose*, which is like a short drill collar with a rounded nose and a hole in the end for mud to exit. The bullnose guides the hole opener along the original hole and minimizes the chances of cutting another hole (Fig. 3-11).

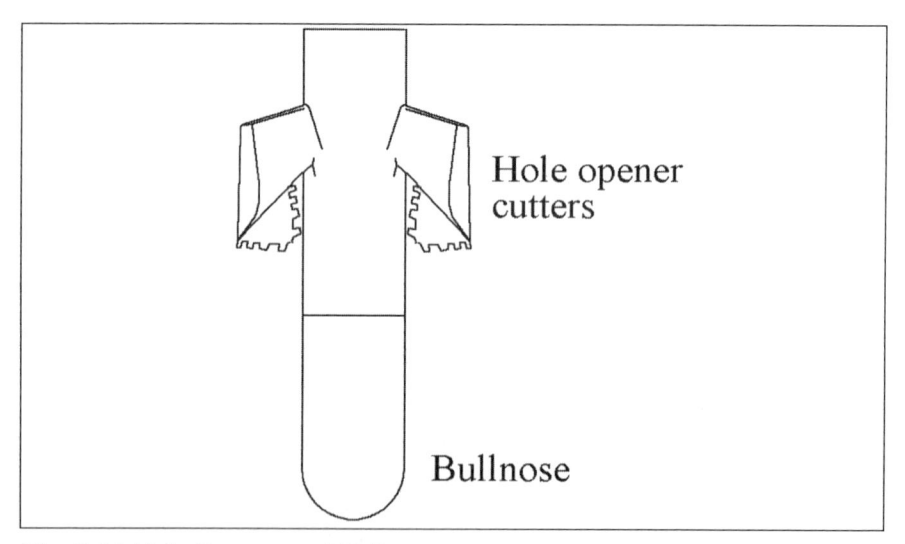

Fig. 3-11 Hole Opener and Bullnose

Running and cementing surface casing

Now a 26" diameter surface hole is drilled below the conductor to a depth that gives sufficient formation strength to contain any likely overpressures if a kick occurs in the next hole section. The casing is cemented in place with the casing shoe in this strong formation.

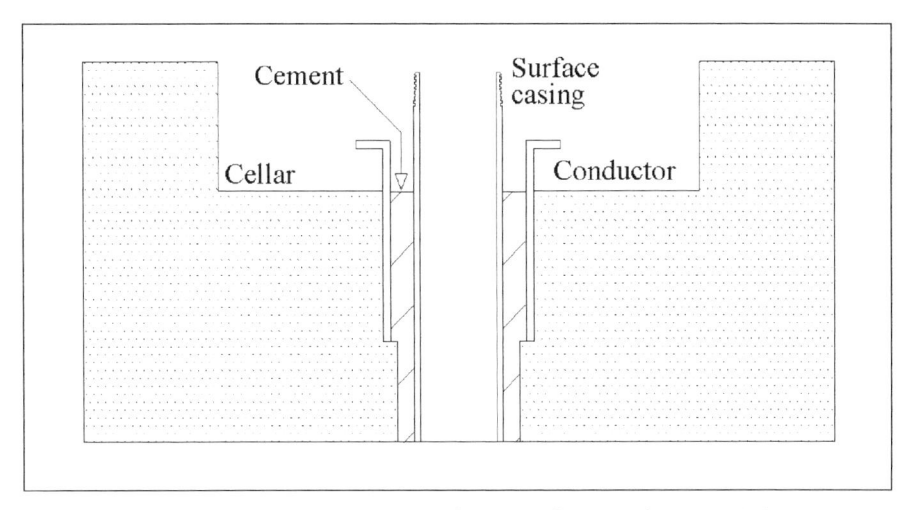

Fig. 3-12 Situation at the End of the Surface Casing Cement Job

Once the cement is hard, the top joint of casing is unscrewed. The connection just above the conductor top is left a little less tight and now it can be unscrewed by turning the top joint of casing, which is sticking up above the rotary table. Once this is out of the well, the diverter is removed, leaving the top of the casing exposed. If for some reason it can't be unscrewed, the diverter can be lifted a little bit and the casing cut above the connection with an acetylene torch. The diverter cannot be removed until this top joint of casing is out (Fig. 3-12).

Attaching and testing the BOP

At the top of the surface casing is a screw thread. A special piece of equipment, called a *casing head housing*, is screwed on to this. The housing has a flange on top, which is used to attach the BOP. In addition, this spool will support the weight of the next casing. The diagram below shows how this works; a casing hanger is screwed onto the top of the next string of casing. This casing hanger sits in the casinghead housing and supports the weight of the intermediate casing string (Fig. 3-13).

If for some reason the thread on top of the surface casing cannot be used, a "weld-on casinghead housing" is welded to the top of the casing.

Once the casinghead housing is in place (whether a screw-on or a weld-on type), the BOP can be positioned and attached to the top of the casinghead housing.

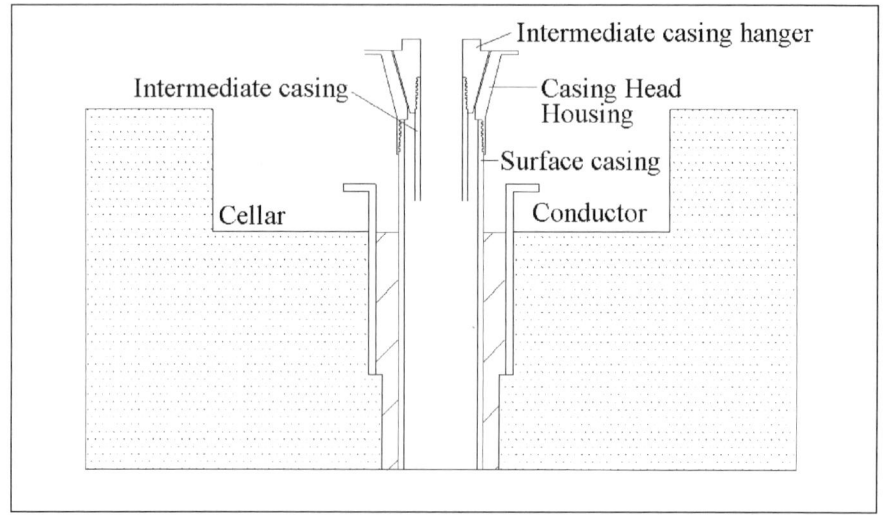

Drilling Technology in Nontechnical Language

Fig. 3-13 Intermediate Casing Suspended in the Casinghead Housing

The BOP equipment and control system must be tested to insure that everything works. It's also vital to pressure test the entire BOP and each part of it to insure that it will hold the pressure that it should. BOP systems come in standard pressure ratings of 2000 psi, 3000 psi, 5000 psi, 10,000 psi, and 15,000 psi. BOPs also come in different sizes, the size being the nominal inside diameter of the BOP.

The time to attach the casinghead housing and to attach and test the BOP on an average operation would take around 10 to 15 hours, depending on the problems encountered. While the BOP is nippled[17] up a survey tool is normally run inside the casing on wireline—this will measure the direction (called *azimuth*) and inclination of the wellbore every 30 feet or so. Knowing the azimuth and inclination at various depths in the well allows the actual path of the wellbore to be calculated. This is necessary to know exactly where the bottom of the casing lies.

Drilling out the casing and testing the formation strength

The surface casing used in this example well has an outside diameter of 20" and an inside diameter of 19". The burst strength (the internal pressure that the casing will hold without breaking) of this particular casing is 2410 psi.

Chapter 3 • Planning and Drilling an Exploration Well on Land

At the bottom of the casing is the *float shoe* (a valve made of cement with some plastic components). When the 17-1/2" diameter drill bit is used to continue drilling, the first thing it will encounter is the top of the float shoe. This, and any cement below it, is drilled until the drilled depth of the previous bit is reached. After drilling to the original depth, and drilling some virgin formation, the strength of the formation below the casing shoe has to be tested. Usually the well is first displaced to a new mud system and this ensures that the density of all the mud in the well is accurately known.

The test is done with some variation of the following procedure. Drill out through the float shoe with the old "spud mud" system. Once the float shoe and cement are drilled, start to pump the new mud while drilling around 15 feet of new formation. As the old mud comes out of the annulus, it is normally dumped into the waste pit. After drilling the new formation, keep circulating new mud until all of the old mud and all drilled cuttings are circulated out. When clean mud returns out of the annulus at the same density as the mud going in, a uniform fluid of known gradient fills the well.

The drill bit is pulled back until the bit is inside the surface casing. The BOP is closed so that it forms a seal around the drillpipe. This leaves the situation drawn in Figure 3-14.

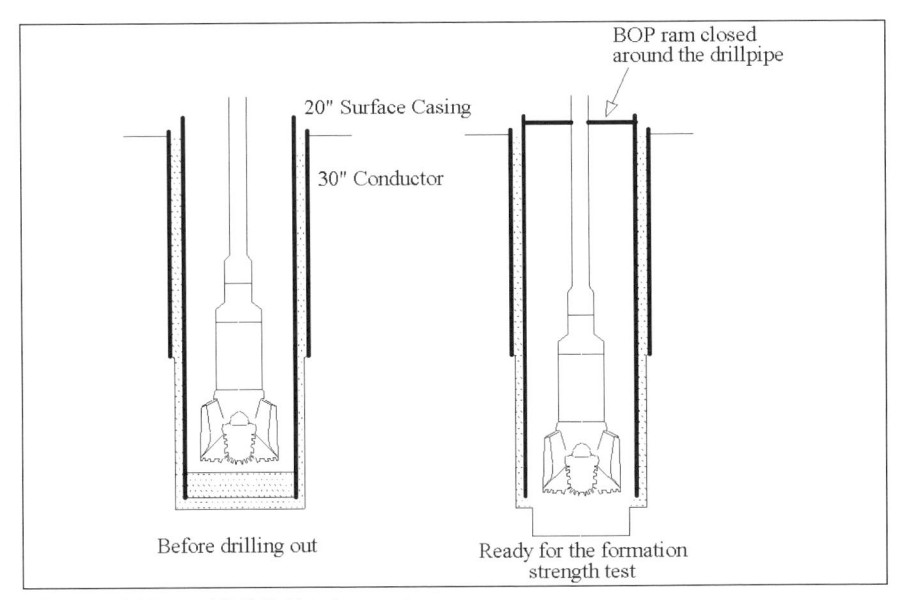

Fig. 3-14 Use of BOP Equipment

Drilling Technology in Nontechnical Language

Fluid is now slowly pumped into the well through the drillpipe. As the annulus is sealed at the top by the BOP, the well becomes pressurized because the fluid has no way out. This pressure is recorded at the surface. Fluid is pumped into the well a bit at a time (around 10 gallons) and each time, the well is left for the pressure to stabilize (become steady) for 2 or 3 minutes. At first, the relationship between the volume pumped and the resulting pressure is proportional—for example, for every 10 gallons pumped into the well, the pressure increases by 50 psi. After pumping 40 gallons, the pressure is 200 psi. Eventually, pumping in the same additional volume (another 10 gallons) results in a lower than expected rise in pressure (say 30 psi). This point is called the "leak off" because the exposed formation has *just started* to allow fluid to leak into it. If any more fluid is pumped into the well, the formation is likely to completely fracture which will of course reduce its strength.

The actual pressure on the formation is calculated by adding the final surface pressure to the hydrostatic pressure of the fluid in the well. Let's assume that the fluid in the well has a gradient of 0.5 psi/ft and it is 1000 feet vertically deep, so the hydrostatic pressure exerted at the bottom of the well is 0.5 x 1000 = 500 psi. If the final surface pressure was 230 psi then the total pressure on the formation downhole is 730 psi. This pressure must not be exceeded (and preferably not even approached closely) because if the formation fractures under pressure, fluids might travel outside the casing to the surface. If the load bearing capacity of the surface soil is reduced, that could cause the rig to collapse and, of course, is likely to cost lives.

Knowing the depth of the formation and the pressure it will bear, the formation strength gradient in psi/ft can be calculated. In this example it will be pressure divided by depth, or 730 ÷ 1000 = 0.73 psi/ft. Now it is possible to calculate a figure that the driller[18] on the rig must *always* know—the maximum surface pressure that can be exerted on the well with the particular drilling mud density in the hole. It's called the maximum allowable annular surface pressure (MAASP) and is very easily calculated as follows:

MAASP = *(formation strength gradient − mud gradient) x vertical depth*

With the 0.5 psi/ft mud in the hole, the MAASP will be 230 psi (which was the leak off pressure at the end of the test). However if the mud gradient is increased later on while drilling, the MAASP will reduce. So at a mud density gradient of 0.6 psi/ft, MAASP must be recalculated and will be

$$(0.73 - 0.6) \times 1000 = 130 \text{ psi}$$

If the well drills into an overpressured formation and takes a kick, the maximum pressure on the annulus at the surface is 130 psi. If this pressure is exceeded, then the formation just under the shoe is likely to fracture.

Drilling the first intermediate hole section

After drilling out the surface casing shoe and 15 feet of new formation, the formation strength was tested. The well was circulated[19] to a new mud system.

The next chapter will discuss drilling a well accurately to a target that is not directly below the surface location. However, the example exploration well here is a vertical well (with the target directly under the rig) so there are no special considerations for making the wellpath follow a particular course to the target. It is just required to be reasonably vertical and straight.

A well can be nearly vertical but might have a spiral path downwards. Even a slightly spiral well can cause problems and this situation is best avoided. A drillstring under high tension (from the weight of the steel below it) in a spiral hole will touch the hole in many places with a high force pressing into the wall. This will cause a lot of friction which will wear the pipe and will make running in and pulling out the drillstring more difficult. It will also make the wellbore less stable, so that it enlarges with all the problems related to an overgauge hole.

One tool in our armory to keep the well straight is the stabilizer.

A stabilizer is run within the bottom hole assembly. Imagine a short section of drill collar, about 5' long, with drill collar connections on top and bottom. In the middle of this collar are fastened blades that stick out, sized so that they touch the wall of the hole. If a stabilizer is run in the hole and if the hole is in gauge[20], the effect is to centralize the drill collars above and below the stabilizer in the wellbore. Now a drill collar, being very thick steel, is quite stiff—it resists bending forces. If a bottomhole assembly is configured with a stabilizer below each of the lowest three drill collars, and as long as the hole is in gauge, then the drill bit requires a very large force

Drilling Technology in Nontechnical Language

to deflect from drilling a straight line. The stabilizers, close to each other and connected by very stiff drill collars, with three points of contact with the wellbore, keep the drill bit drilling in line with the existing wellbore (Fig. 3-15).

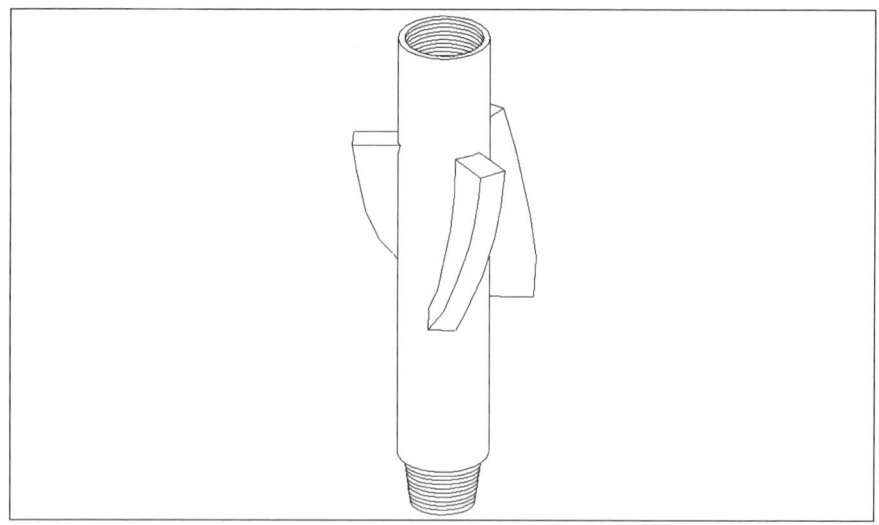

Fig. 3-15 Stabilizer—Conceptual Drawing

At the start of this intermediate hole section, the well is vertical. It points directly towards the target required by the well design. In order to keep the well straight, at least three stabilizers with blades of the same OD as the drill bit are placed close together in the bottom part of the BHA. This BHA design may be called a "packed", "locked", or "tangent" assembly. Other BHA designs are discussed later.

Running this design of BHA also has some drilling advantages. If the BHA included no stabilizers, then if a lot of weight is placed on the drill bit (to make it drill as fast as possible) the lower part of the BHA will start to buckle. This will give unpredictable results, as the bit is no longer aligned with the center of the hole and is likely to drill a spiral hole or even veer off course. If the BHA buckles while the drillstring is rotating, then significant fatigue damage will occur to the drill collars, which may eventually cause something to break. Using a locked assembly allows the well to be drilled faster as well as straighter. Some people consider that running more stabilizers will give more frictional force against the BHA when rotating and tripping[21] in and out, but in fact drilling a straight hole reduces friction far more than might be increased by having a few bits of steel touching the side of a straight hole.

Now the well can be drilled ahead towards the depth to run the next casing, called the *intermediate* casing because it comes after the surface casing, but before production casing. A deep well might have more than one intermediate string, or it may contain none if it's a shallow well where the reservoir can be reached with only one casing string after surface casing.

Once casing point is reached, after one or several bit runs, the bit is pulled out of the hole. Electric logs will most likely be run at this time.

Logging

Logs are discussed in detail in chapter 10.

Tools are run into the hole on special steel wireline, usually 9/16" in diameter. Inside the wireline are electrical wires that connect the tool to a computer unit on the rig. Logs measure physical and chemical characteristics of the formations. In an exploration well, the well is drilled to gain information and most of this information will be obtained by using logs.

This hole section is an intermediate hole section; the reservoir is not yet penetrated. Even though the reservoir is not exposed by the wellbore, there is a lot of information that can be used to help improve the well design and drilling program for the next well. Many people have an interest in the log results—drilling engineers, geologists, geophysicists, and other specialists.

Logging is quite expensive, not just in the cost of the rig time taken, but the logs themselves can be very costly to run. Some logging tools cost in excess of a million dollars, so you don't want to lose one down the hole!

Logs run in the intermediate hole section would typically try to establish

- Formation tops and thicknesses
- Formation lithology, porosity, pore fluids salinity, and presence of any hydrocarbons
- Continuous hole profile, showing the diameter in two directions (so that an overgauge hole that is enlarged more in one direction than the other can be recognized)

- Formation compressive strengths (measured using sound waves, useful when deciding what drill bits to use on the next well)

Running and cementing the first intermediate casing

If there were any hole problems while logging, the drilling BHA and drill bit last used may be used to go to the bottom of the hole and back. This can check if there is any debris on the bottom (rock material that remained in the hole, or cavings that have come from an unstable formation since the last trip out). The action of a round trip with a bit and BHA can be beneficial; if there are any tight spots while tripping in (places which require some force to pass), the driller can start pumping mud and rotating the drillstring to "ream out" the tight spots which might otherwise cause a problem when running the casing. This round trip is called a *check trip*.

On some operations, the rig crew run in with the casing, on other operations a specialist contractor will send out men and equipment to work with the rig crews.

As with surface casing, a float shoe is screwed onto the bottom of the casing. After two joints of casing (about 80 feet of casing higher) a *float collar* is run. A float collar is similar to a float shoe, except it has a thread at the bottom end so that casing can be run underneath it. This gives two float valves in the casing string, so that if one fails, the other acts as a back up.

The casing is run into the well. During running, the crew will put a hose into the casing every two or three joints and fill it up with mud (because the float valves on the bottom prevent mud from entering the casing at the bottom). Now the casing outside diameter is smaller than the drill bit, but it is much less flexible than the drilling assembly. It is possible to have problems running casing into the hole because of this and because of the large surface area of the casing. If there is a part of the hole that is enlarged, and this is just above a different rock formation that is in gauge, there will be a "ledge" in the hole. The casing might hit this ledge and it might be difficult to pass through it. There is a better chance of getting stuck with casing because it is very large and rigid when compared to the drillstring.

Once the entire casing is in the well, a casing hanger is screwed onto the top joint of casing. This has a cone shaped profile that fits into the casing head housing on top of the surface casing. Figure 3-16 shows the situation before the intermediate casing has been lowered down so that the casing hanger lands inside the casing head housing.

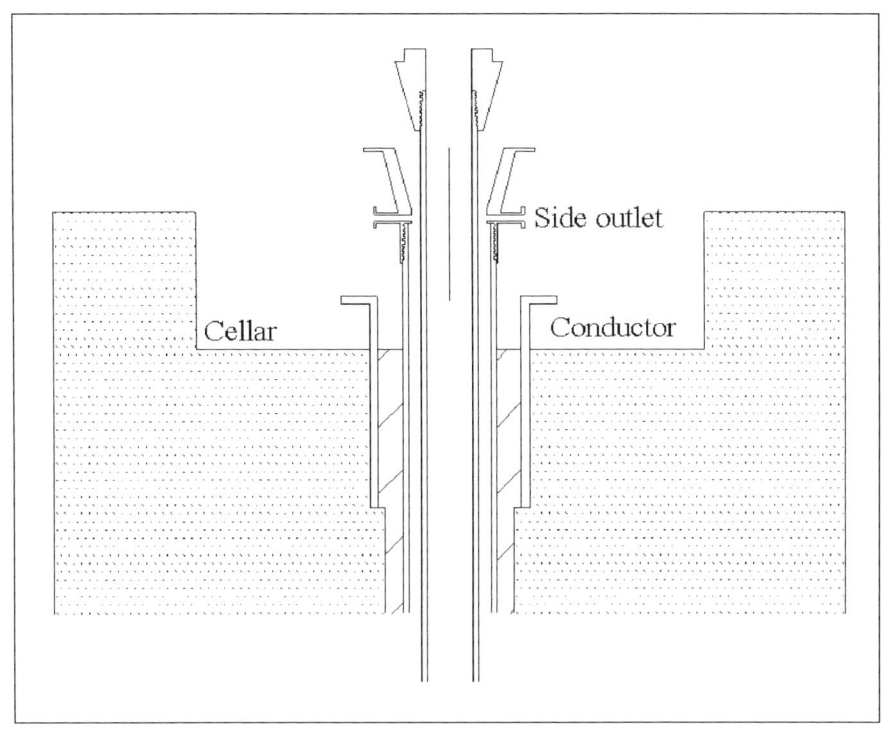

Chapter 3 • Planning and Drilling an Exploration Well on Land

Fig. 3-16 Intermediate Casing Before Landing in the Casinghead Housing

Once the casing hanger has landed in the profile inside the casinghead housing, seals on the outside of the hanger create a pressure tight seal between the hanger and the housing. Shown on the drawing above are pipes leading out from the casinghead housing—these are called *side outlets*. After the casinghead housing is attached to the surface casing, valves are attached to these side outlets. Now when landing the intermediate casing, these valves are open to allow mud to flow from the well when pumping cement down the casing. Hoses are attached to the outside of these valves and the hoses take returns from the well to the rig mud tank system. This is illustrated in Figure 3-17:

With the casing landed and the hoses attached, mud is circulated down the casing. Gradually the flow rate is increased while watching the tank levels very carefully to detect the start of any mud loss downhole. Increasing the flow rate also increases the pressure in the annulus and eventually loss to a formation downhole might occur. The annulus between the casing and the hole is much smaller than it was with the drillstring and this

Drilling Technology in Nontechnical Language

gives higher pressures while circulating. As soon as mud loss is detected, the pump is slowed down a little. This flow rate will be the maximum flow rate during the cement job. Circulating continues until about 120% of the casing's inside volume is pumped. This insures that there is no debris (such as rags, brushes, or somebody's hard hat) inside the casing, which can plug the float valves—which would be disastrous if it were to happen while cementing. At the same time, this helps to improve the cement job for reasons that will be discussed in chapter 9.

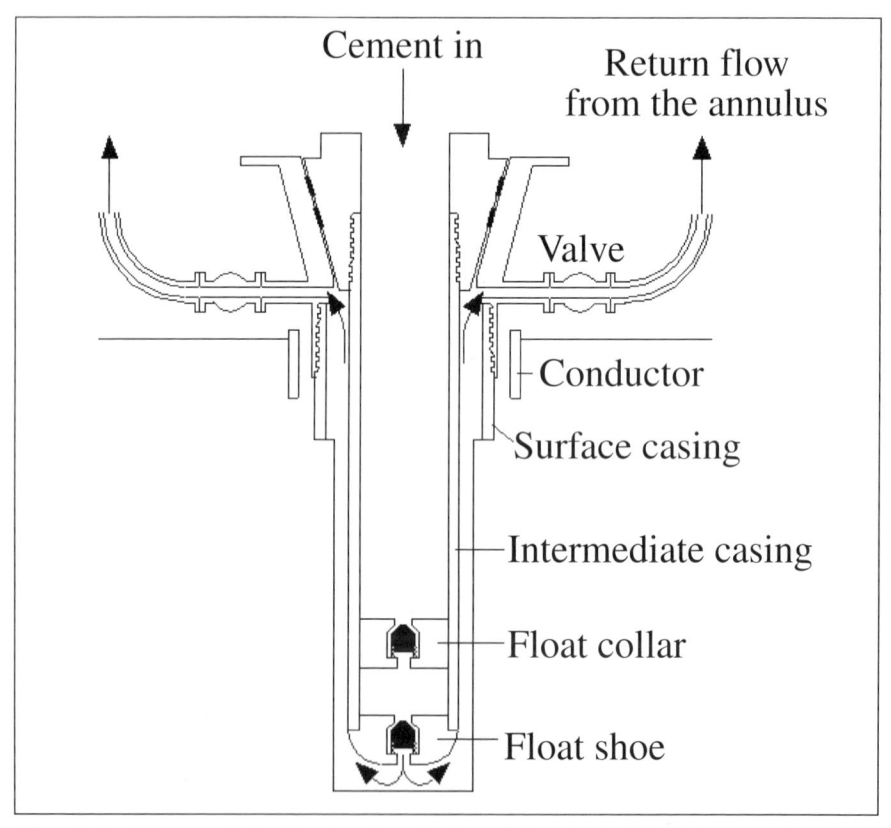

Fig. 3-17 Intermediate Casing Landed and Ready to Start Cementing

Once satisfied that all is well, cement is pumped down the casing. In front of the cement is a rubber (or plastic) plug that seals in the casing. This moves down ahead of the cement, wipes all mud off the inside of the casing, and separates the mud and cement to avoid contamination. This is

Chapter 3 • Planning and Drilling an Exploration Well on Land

called the *bottom plug*. The bottom plug has a diaphragm on the top that will break when the plug hits the float collar, allowing cement to flow through the plug and down through the float valves.

Behind the cement, another plug is used that is similar to the bottom plug, but is solid and does not rupture when it lands on the bottom plug. This second plug is called the *top plug*. It's job is to wipe cement from the inside of the casing and to separate the cement from the mud behind it. When the top plug lands on the bottom plug, they seal and prevent further movement of fluid down the casing. This allows the driller to see exactly when the cement is in place because his pumping pressure will increase. At this time, the driller can pressure test the inside of the casing to make sure there are no leaks in it.

After pressure testing above the casing hanger to insure that the hanger seals work, the BOP can be removed. Then, a housing called a *casing spool* is added to the wellhead, in which the next string of casing will land.

The assembly of casings, hangers, and spools is called the *wellhead*. The status of the wellhead is illustrated in Figure 3-18.

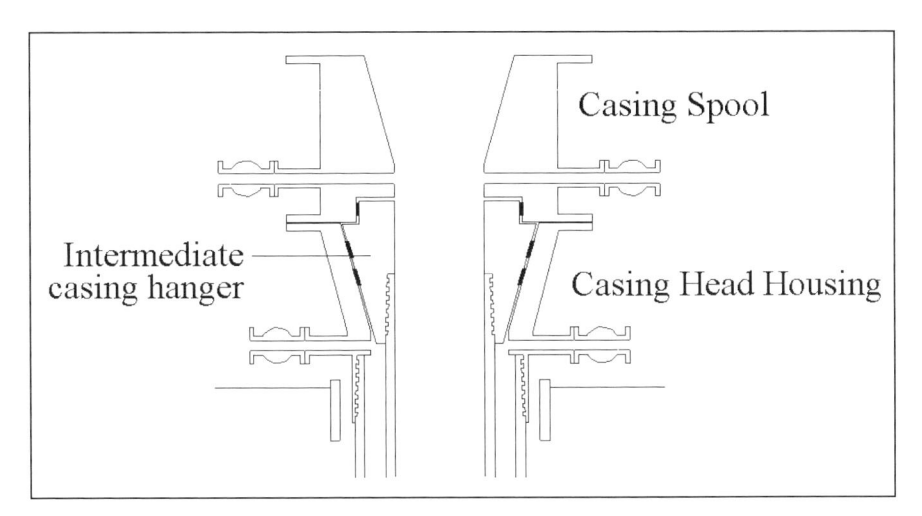

Fig. 3-18 Building Up the Wellhead for the Next Hole Section

The inside of the casing spool has a similar profile to the casinghead housing, only smaller. Each string of casing requires a casing spool to land in (with this particular type of wellhead).

The next BOP will be nippled up on top of this spool. It will be smaller than the first one and will have a higher pressure rating.

Drilling the production hole section

Imagine now that another hole section was drilled and another intermediate casing cemented in place. Things will not be too different from the intermediate section described above. This new casing is set approximately 500 feet above where the top of the reservoir is expected. The plan now is to drill through the reservoir, then roughly 200 feet below it, log the hole and run a liner.

This is the part of the hole that everybody is really interested in. If hydrocarbons are present, then a lot of wireline logs are run over several days. If the wireline logs indicate that hydrocarbons are likely to be in commercial quantities and can be produced, then a liner will be cemented in place and holes blown through the side of the liner to allow the hydrocarbons to be produced during a test.

A *liner* is essentially a string of casing that does not extend all the way to the surface. A liner is suspended from a liner hanger, which uses hardened steel teeth to dig into the last casing ID to suspend the liner. The advantage of running a liner is the reduced cost (a much shorter length of casing pipe, less cement needed, etc.). The disadvantage is the increased complexity because of the tools that must be manipulated while deep inside the well (Fig. 3-19).

While drilling through the reservoir section, there are indications on the surface of hydrocarbon presence. When drilling through a gas-bearing zone, increased levels of gas dissolved in the drilling mud can be detected. When drilling through oil-bearing rock, the rock cuttings will show the presence of oil in the pore spaces.

Now it is very important to gain the maximum amount of information from the reservoir. The best way to assure this is to take a core sample while drilling through the reservoir. Often on an exploration well, the coring equipment will stand by on the rig until the first good indication of hydrocarbon presence, where the drill bit will be pulled out and the drilling BHA replaced with a coring assembly.

A core is taken by drilling with a special bit (called a *core bit*) which has a hole in the middle. As the bit drills a doughnut shaped hole, a column of uncut rock will stick up inside the bit. Behind the bit is a special

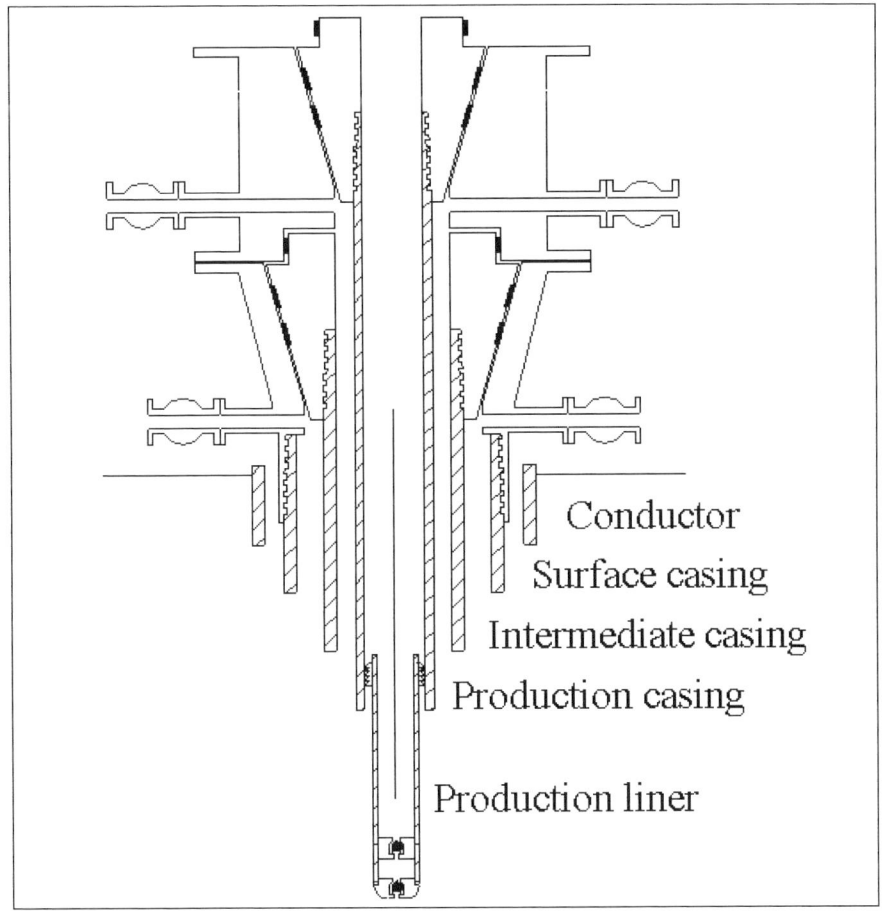

Chapter 3 • Planning and Drilling an Exploration Well on Land

Fig. 3-19 Final Well Profile with the Production Casing and Liner in Place

mechanism for gripping this rock and holding it in a special container (Fig. 3-20).

Coring is slow and expensive but the value of the information usually makes this worth while because it allows better decisions in the short term (designing the well test; discussed in chapter 10) and in the long term (should the reservoir be developed and how?).

Once the reservoir is cored (which may take more than one core bit run to achieve) the driller will run in with a normal drill bit, ream through the cored section and continue drilling to total depth (TD).

Logging the hole

The same logs are used here as per the previous hole section. In addition, some extra logs will help evaluate the reservoir and to yield information needed to design the well test program.

One tool that will almost certainly be run, places a probe against the formation and opens up a sample chamber on the inside of the tool. This can take a sample of formation fluids at the composition that is downhole for analysis at the surface. These tools can also take very accurate pressure readings and can measure the formation permeability at the wellbore face.

Fig. 3-20 Model of a PDC Core Bit Seen from Below. Note the hole in the center for the core sample.

Running and cementing the production liner

A liner, being effectively a string of casing with the top half missing, is assembled and then run in the hole below a string of drillpipe. At the top of the liner is a "liner hanger", which may be hydraulically or mechanically manipulated to move hardened steel elements out to grip the inside of the previous casing. Cement is pumped down the drillstring, with special plugs to separate the fluids. Once the cement is in place outside the liner, the drillpipe is released from the liner hanger and pulled back to surface (Fig. 3-19).

Production Testing the Well

The cuttings, logs, and core samples show that there is gas and that there is likely to be a commercially worthwhile amount of oil. There exists sufficient permeability for hydrocarbons to be produced into the wellbore. The inside of the well needs to be cleaned out, a string of tubing run in the well, holes made through the liner into the reservoir (called *perforating*), and the well must flow to test the reservoir's characteristics.

Preparing the well for the test string

The inside of the well contains drilling fluid that has a high level of solid materials suspended in it. These solids may be part of the mud system, drilled solids, rust products from the steel in the well, and other debris. If the liner is perforated with this fluid in the well, those solid particles will enter the perforation tunnel and can plug off pore spaces in the reservoir. This is one possible mechanism that will reduce the ability of the well to produce oil. This damage must be minimized.

First, a drill bit is made up on the BHA, followed by a tool called a *scraper*. A scraper has sets of spring loaded blades which push against the casing, scraping off loose rust, grease, etc. When this is at the bottom of the liner, clean, filtered brine is pumped around the well, which contains very little solids. This pushes the drilling mud out of the hole. With the mud out, brine is circulated around until it is as clean as is specified by the test program. When the well is as clean as it needs to be, the bit and scraper are removed from the well.

Running the test string

Now the well is clean and contains a clean, specially formulated brine that is free of solids. A string of tubing is run, through which the hydrocarbons will be produced. In the "old days" drillpipe was often used and hence the production test became known as a "drill stem test". Drillpipe is not used any more for well testing because there is no guarantee that it will seal with high-pressure hydrocarbons inside it.

The well should not produce hydrocarbons from inside the casing except through a separate tubing string. If tubing is in place, the well integrity is much better (safer) and the well may be controlled by circulating fluids around, which cannot be done without tubing in the well.

The tubing string will have various tools incorporated. One will be a sub surface safety valve (SSSV). The SSSV is positioned some hundreds of feet below the surface (or below the seabed on an offshore well). It is a valve that is normally closed, but is held open by pressure on a special line that is run to the surface outside of the tubing. If something goes wrong the con-

Drilling Technology in Nontechnical Language

trol line to the SSSV is depressurized and this closes the valve. In addition, if something disastrous happens, such as the wellhead gets blown off by marauding Iraqi troops, then the well will shut itself in automatically. Had the wells in the Kuwaiti desert incorporated SSSVs, the result of Saddam Hussein's deliberate sabotage would have been minor oil spillage around the wellheads and not the environmental disaster that he wanted and obtained. As it was, SSSVs were unfortunately not used (probably to save a few dollars).

The tubing is suspended in the well at the wellhead, just as the casings were suspended at surface. After running the last string of casing, the BOP was removed and a tubing head spool placed on the wellhead. This goes on top of the casing spool in the same way that the casing spool went on top of the casinghead housing. With the complete tubing string in the well, a tubing hanger is screwed on top of the tubing and this lands inside the tubing head spool. Special bolts in the spool are screwed in to lock the tubing hanger in place. After pressure testing to make sure the seals all work, the BOP is removed and a special assembly of valves, called the *Christmas tree*, is bolted on top of the tubing head spool (Fig. 3-22).

Now the well is ready to perforate. In preparation for perforating, a special BOP is placed on top of the Christmas tree, which will allow the well to be perforated with wireline while having the possibility to seal around the wireline.

Perforating the well

The logging cable is now used to run perforating guns inside the tubing. A perforating gun consists of some sort of carrier mechanism, inside of which is a set of shaped explosive charges. These are run in the well so that they are opposite the reservoir. When they are detonated, the shaped charge efficiently focuses the explosive energy in a narrow path outward, developing pressures of millions of psi within a jet of plasma. This punctures the liner and creates a tunnel through the rock, often up to 22" in length.

Testing the well

Well testing is discussed in more detail in chapter 10.

The hydrocarbons produced during testing are normally burned off. There is nowhere that the fluids can be safely stored and in the case of this exploration well in the desert, no infrastructure exists which could capture the produced oil and get it to a refinery.

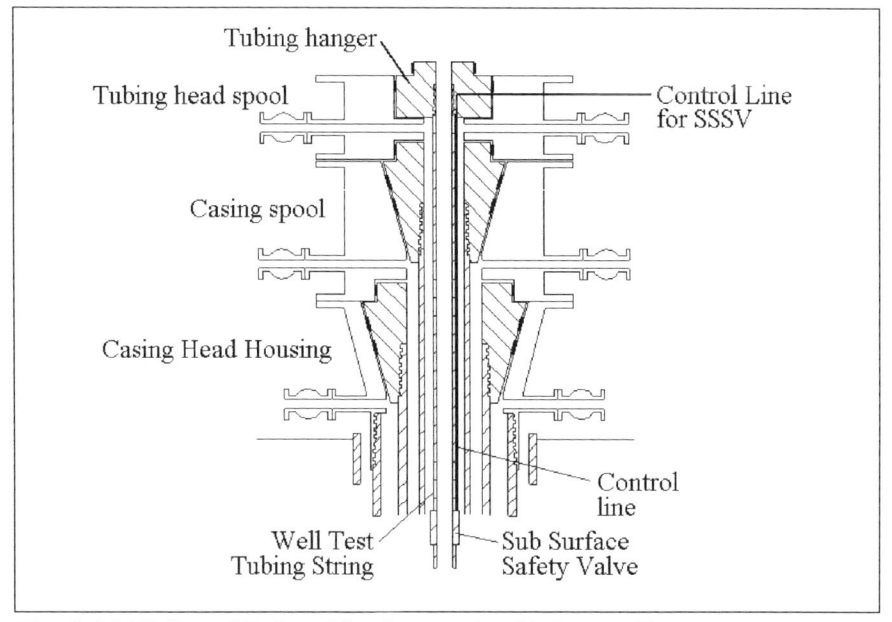

Chapter 3 • *Planning and Drilling an Exploration Well on Land*

Fig. 3-21 Wellhead Before Nippling up the Christmas Tree

Killing the well after the production test is complete

Once the well test is over, the well will be "live" from the pressure inside the tubing. The well is "killed" by pumping heavy fluid down the tubing, which will build up enough hydrostatic pressure to overcome the reservoir pressure. Once the well is completely dead, the well can be abandoned.

Abandoning the Well

A well is abandoned when there is no conceivable use for it after the well test. Well abandonment involves placing cement in the well to prevent reservoir fluids leaving the reservoir and moving up towards the surface. After isolating any hydrocarbon or freshwater zones with cement, the wellhead is removed.

Removing the test string

Once the well is dead, plugs are placed inside the tubing to ensure that, even if pressure does start to build up inside the tubing, nothing can come back to surface. The Christmas tree can then be removed and the BOP nippled up on top of the tubing head spool.

The bolts holding the tubing hanger in place are removed and the test string can be pulled out.

Making the well safe into the future

The first task is to run in with some pipe and pump cement down so that it is left around the perforations. Often, special plugs will be run on wireline electric line to give additional security.

Special cutting tools are available which can cut the casing. The casings will be cut above the top of cement outside the casing so that the upper bit of casing can be pulled out of the well. Finally the conductor is cut, cement dumped on top and the site restored for future non-drilling use.

Chapter Summary

This chapter examined the processes, equipment, and decisions involved in drilling an exploration well in the desert. Many new technical terms and concepts were introduced and explained.

All of the explained terms are also in the chapter Glossaries.

Glossary

[1] **Operator.** An oil company whose business is finding, producing, and selling hydrocarbons. Well-known examples are Shell, Exxon, and British Petroleum. There are also smaller, (independent) operators who may be very small in terms of market capitalization (a few tens of millions of dollars).

[2] **Wildcat well.** A well drilled in an area where no drilling was done before. A wildcat well is much more risky than one drilled in an area that has had some exploration wells drilled and subsurface conditions are not totally unknown.

[3] **Abandon a well.** A well is abandoned by pumping cement into the well to plug off all permeable zones. The casing strings are cut somewhere below the surface and cement placed on top of the cut casings. The surface (on land) will be restored with soil, the seabed offshore will naturally fill up with sediments.

Chapter 3 • Planning and Drilling an Exploration Well on Land

[4] **Gas-oil contact.** The depth of the top of the oil/bottom of the gas in the reservoir. Oil-water contact—similarly, the depth of the top of the water/bottom of the oil in the reservoir.

[5] **Core samples.** It is possible to drill with a special bit that has a hole in the center. A doughnut shaped hole is drilled with a column of rock sticking up inside the drill bit. This can be broken off and recovered to the surface to give a sample, or "core", of formation. Core samples, properly recovered and handled, can give a wealth of valuable information about the reservoir that can then help to optimize the plan for developing the field. Cores can be cut in almost any type of rock.

[6] **Internal reservoir characteristics.** A well is tested by allowing it to flow, then closing the well for a time, then opening and closing it for specified times and at different flow rates while flowing. The pressures measured while flowing and while closed can give a lot of information about the reservoir. Well testing is a specialized and very interesting operation requiring high levels of expertise, extremely accurate and precise instruments, and powerful computer programs. Chapter 10 goes into a bit more detail. Suggested references are given at the end of the book for further reading.

[7] **Mechanical skin.** This is a term given to the damage done to the reservoir in the region close to the wellbore. Imagine that if many of the pore spaces become plugged by solid materials present in the mud, this will form a kind of skin on the reservoir, which will make it less permeable, and will therefore reduce production potential. Well testing can measure the extent of this skin by flowing at different flow rates and measuring how quickly the wellbore pressure rises when the well is closed in.

[8] **Shallow gas**. A gas deposit that is encountered before setting the surface casing.

[9] **Seismic survey.** Sound waves are sent from the surface and the reflected sounds recorded. These reflections can be analyzed to show details of subsurface structures.

[10] **Diverter.** A special type of blowout preventer that diverts fluids flowing from the well away from the rig.

Drilling Technology in Nontechnical Language

[11] **Lead time.** The time it takes from ordering something to it being available for use.

[12] **Drillstring.** The pipes and tools that are run into the hole in order to drill. The main components of a drillstring are described later, including the drill bit, drill collars (heavy pipe used to push down on the bit), and drillpipe (lighter, thin walled pipe used to run the drill bit and drill collars to the bottom of the hole).

[13] **Spudding the well.** The act of starting to drill the well is called "spudding".

[14] **Barrels.** A measure of volume. One barrel equals 38 Imperial gallons or 42 U.S. gallons. Normally in the oilfield, gallons refer to U.S. gallons and not Imperial gallons. Barrels are very commonly used to discuss fluid volumes while drilling. Barrels are also commonly used to refer to the volume of oil produced from a well. Barrels are normally abbreviated "bbls".

[15] **Returns.** A reference to the fluid leaving the well at the top of the annulus. Fluid "returning" to the surface.

[16] **Lands.** Describes when one piece of equipment is placed on another, such that the lower piece of equipment supports the weight of the upper. So a tool "lands" on a landing ring. This expression applies also to casing, which is lowered into the hole and "lands" on a conical profile inside a large steel bowl.

[17] **Nipple up.** This expression refers to a process of assembling equipment. Conversely, "nipple down" refers to disassembling equipment.

[18] **Driller.** The person in charge of a drill crew. Most rigs work 24 hours a day, using two drill crews working 12 hour shifts. The driller leads a drill crew and he is also responsible for ensuring that a safe and efficient drilling operation is carried out on the drill floor. The drillers are key people on the rig and their decisions have a huge influence on the safety and efficiency of the operation.

[19] **Circulate the well.** The act of pumping a fluid all the way around the well, from the surface to the bit, up the annulus, and back to the surface.

Chapter 3 • Planning and Drilling an Exploration Well on Land

[20] **In gauge.** The same diameter as the drill bit. An overgauge hole is larger than the drill bit diameter and an undergauge hole is smaller than the drill bit diameter.

[21] **Tripping.** To lower pipe in to the hole ("trip in"), pull it out of the hole ("trip out"), or to pull out, change something (like replace the drill bit), and run back in again ("round trip").

[21] **Brine.** A solution of salts in water. Many different salts are used in formulating brines, depending on the chemical and physical characteristics required.

Chapter 4

Planning and Drilling a Development Well Offshore

Chapter Overview

This chapter will explain how a development well might be drilled offshore through a template on the seabed. A development well is drilled to produce hydrocarbons when a field is commercially exploited.

The example well will drill a horizontal wellbore in a reservoir containing oil, water, and gas to efficiently produce oil.

Concepts from the first three chapters will be referred to, but not redundantly explained. New concepts specific to the type of well and rig under discussion will be introduced and explained.

Well Planning

In modern developments, wells positioned horizontally in the reservoir are becoming fairly common. Horizontal wells are often drilled to follow the upper part of the reservoir (in a gas well) or to stay a certain distance above the oil-water contact (in an oil well). This remarkable feat will be covered later in the book.

Coring is not often called for in development wells, unless there is some remaining uncertainty. The logging program will also be less extensive (and less expensive!) than in an exploration well.

As each well is drilled, it will be left with a special cap on top—a *suspension cap*. This protects the wellhead from debris and corrosion until the platform is in position over the template (Fig. 4-1).

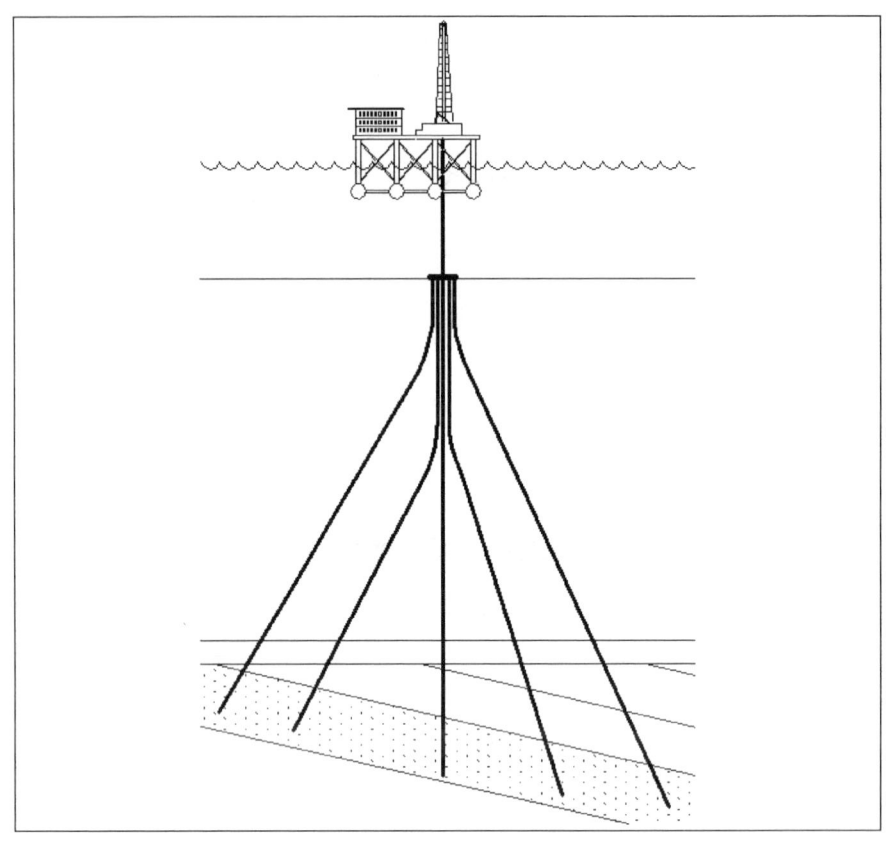

Drilling Technology in Nontechnical Language

Fig. 4-1 Directional Wells Drilled from a Floating Rig Through a Template

The completion design for a development well must ensure that the well can be produced safely, efficiently, and economically. The production engineers will consider various aspects of producing the well when designing the completion. In particular, attention will be paid to the following:

1. **The well inflow configuration.** This governs how hydrocarbons flow from the reservoir into the well. Sometimes it is necessary to pressurize the well so as to create fractures in the reservoir, increasing permeability towards the wellbore. Acid might be used to create channels in the rock and increase permeability around the wellbore. These activities are termed *stimulation*.

2. **The well outflow configuration.** This governs how the hydrocarbons flow from the bottom of the well to the surface. This will dictate the size and type of tubing used, as well as accessories run as part of the tubing string (such as downhole valves or pumps) (Fig. 4-2).

Fig. 4-2 A Sand Control Screen, Cut Away to Show the Screen Elements

If the well gets too close to the oil-water interface, the water (being less viscous than the oil) will "break through" the oil and flow to the well. This is called *coning*. Once coning starts, the well will produce water and not too much oil. Similarly, if the well gets too close to the gas, the gas will be preferentially produced. In a vertical well, there is only a short interval of the well that can be perforated to produce the oil. This means that to produce a reasonable amount of oil, the speed that oil has to flow towards those few perforations is high, which will tend to draw gas from above or water from below towards the well. With a horizontal well, the contact area between the reservoir and well is large, so flow rates in the reservoir towards the well will be low, which will reduce the tendency of gas or water to break through. This is one reason for drilling horizontal wells (Fig. 4-3).

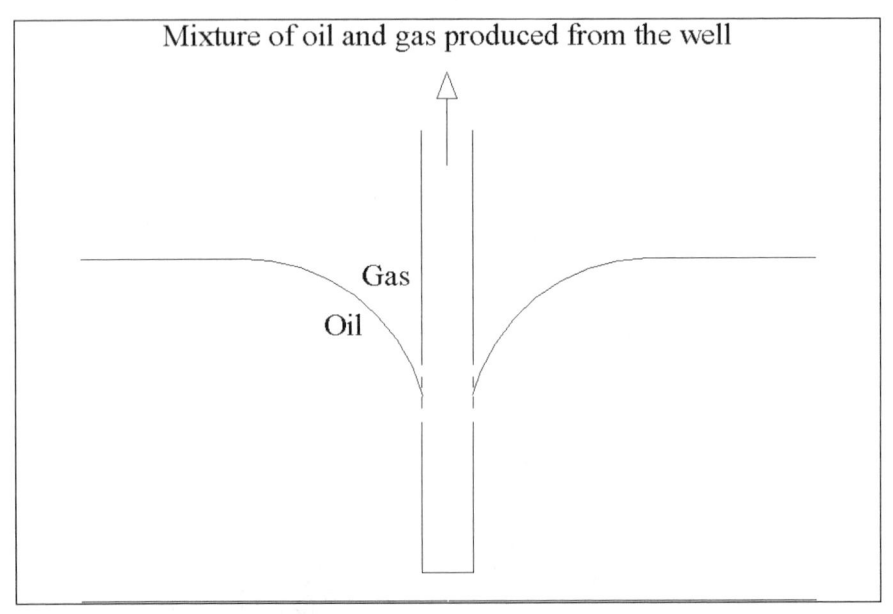

Fig. 4-3 Coning—Gas Breaking Through Oil

Completion design is normally done by the production engineers. However, the completion design affects almost every aspect of the well design, so the drilling engineer must know, at least conceptually, what the completion will look like.

Hole and Casing Sizes

With a development well, the contingency to run an extra string of casing if unexpected hole problems are encountered is less important than on an exploration well. If a well can be drilled one size smaller (20" conductor instead of 30") then generally it will be cheaper to drill the well by around 15%. There is a strong incentive to avoid building in unnecessary contingencies against remote possibilities when the cost penalty is so large.

The hole size of the final section is dictated by the required size of the completion tubing and also by the type of completion. In a traditional completion, the completion tubing extends all the way down to the reservoir, inside the production casing or liner. A completion can also be designed using a liner as part of the completion tubing. One advantage is that the well can be drilled using smaller hole sizes saving a lot of money. If the production of the well requires 5" diameter tubing then the well could be finished by drilling a 6" hole for a 5" liner rather than by drilling an 8-1/2" hole for a 7" liner. This is illustrated in Figure 4-4.

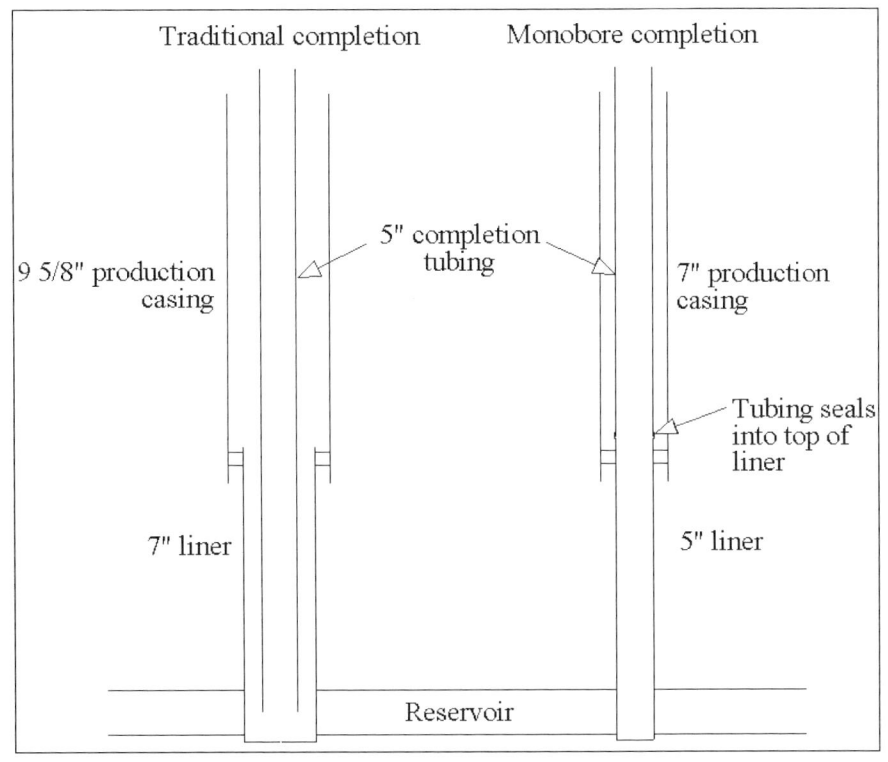

Fig. 4-4 Traditional vs. Monobore Completion

Drilling Technology in Nontechnical Language

Directional profile of the well

The target location is not directly underneath the rig, so the well must be drilled along an accurate path to the target. Once in the reservoir, the well must remain a certain distance above the oil-water contact but not so far above it that it approaches the gas-oil contact. The BHA navigates through the reservoir by measuring the characteristics of the reservoir while drilling, using logging tools that are constructed inside a drill collar. These techniques are called, logically, *logging while drilling* (LWD). The logging tool for this job measures electrical resistivity—the closer it approaches to water, the lower the measured resistivity. As the reservoir is repeatedly logged during the exploration and appraisal[1] drilling, the engineers have a good picture of how the resistivity varies with depth (Fig. 4-5).

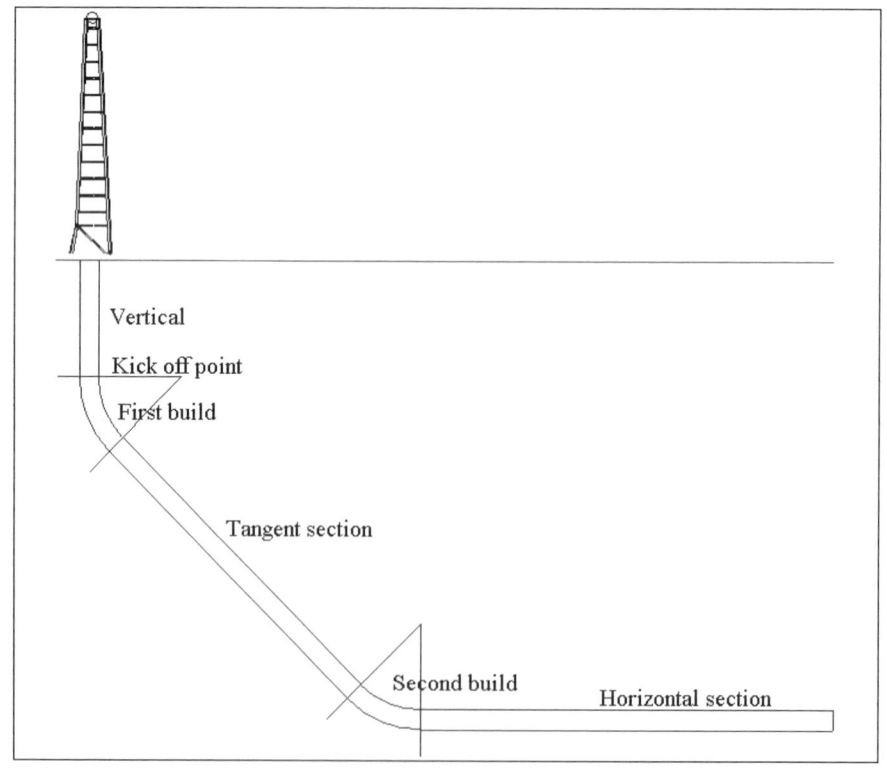

Fig. 4-5 Directional Well Terminology

The well is drilled vertically to begin with. At the kick off point, the rotary drilling assembly[2] is pulled out and a special, "directional" drilling assembly is run in the well. This is designed to exert a side force at the bit, so that the bit starts to drill away from vertical. The direction that the well drills towards is determined by aligning the sideforce in the appropriate direction.

The direction of the well relative to true North is called the "azimuth" and it is usually measured in degrees clockwise. True East will be 90°.

The angle between the wellbore center and vertical is called the *inclination*. The horizontal section of the well, if it is exactly horizontal, will have an inclination of 90°. A vertical well has an inclination of 0°.

There are various tools and techniques used to deviate the wellbore. Directional drilling techniques are covered in more detail in chapter 8.

Writing the well program

With a development well, the engineers have access to a wealth of data from drilling the exploration and appraisal wells. The drilling engineers must learn all the lessons that the previous wells can teach. Of course this is difficult, even if the information is both complete and accurate (which it often isn't). However, there are problems that make this reviewing process even more difficult.

1. Information is often incomplete because people do not always record all the relevant information. In addition, records sometimes go missing for a variety of reasons.

2. In many operations, experience from earlier stages in the project is not properly documented. When engineers leave for another project, their hard won experience goes with them.

3. Most people do not like to admit "mistakes" too readily, so instead of learning from problems, the problems become hidden. In some companies, admitting to mistakes can be promotion or career threatening.

Drilling Technology in Nontechnical Language

Drilling the well

This example well will be drilled from a floating rig. A floating rig moves with the tides, waves, and currents. Floating drilling requires special equipment and techniques to deal with this movement and is discussed in some detail in chapter 5.

Spudding the well and cementing the conductor

A template is a welded steel structure that is placed on the seabed. At each corner, a steel pipe is cemented into the seabed to secure the template. Within the template are placed large pipes with a conical guide above. These are conductor slots and their purpose is to allow conductors to be placed through them (Fig. 4-6).

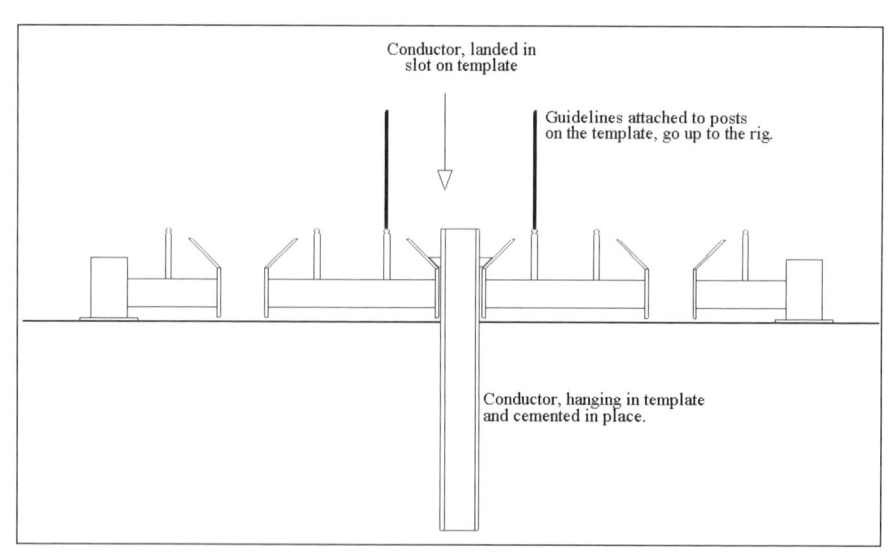

Fig. 4-6 Template Set on the Seabed (side view)

The template has four posts welded around each conductor slot. Guide wires are attached to these posts, extending back up to the rig. These guide wires are used to guide tools in to the well. When a well is spudded, the drilling assembly is loosely tied to the guide wires with 1/2" manila rope. As the bit enters the slot, the rope breaks. As the hole is drilled, spud mud is pumped down the drillstring. The mud and cuttings exit the annulus at the seabed. After pulling out with the drilling assembly, the conductor is run in on drillpipe and cemented in place—similar to the way that a liner is run, as described in the previous chapter. The conductor is guided with rope on the guide wires into the conductor slot.

Drilling for surface casing

With a floating drilling rig, the surface hole is drilled with returns to the seabed. No diverter is run. If shallow gas is encountered, the rig can drop the drillstring and move off location, away from the gas flow. The gas will flow to the sea and the gas plume will disperse as it rises. Any current will move it away from the rig. If a very strong flow was encountered in shallow water, the rig might be endangered by the gas flow destabilizing the rig, causing it to capsize.

As before, the BHA is guided to the template using rope on the guide wires. The surface hole is generally not logged and of course no cutting samples are possible because they are dispersed on the seabed around the template. As this is a development well, the geology is known and so the surface hole terminates at a predetermined depth, where the formation is known to be strong enough to hold pressure if a kick is taken.

At the top of the casing a special tool is screwed on, which contains the cement plugs and allows drill pipe to be screwed on above it. This works in a similar manner to a liner hanger, as described in the previous chapter.

On top of the surface casing is a special "subsea" wellhead housing. This is a section of thick walled pipe. Inside the housing is a profile that the next casing hanger lands on. Subsequent casing hangers stack up on top of each other. On the outside of the wellhead housing is a profile that the BOP latch onto (Fig. 4-7).

Special underwater blowout preventers are set on top of the well after setting surface casing.

After testing the sub-sea BOP on the rig to ensure it functions correctly and holds pressure, large diameter pipe (called *riser* pipe) is connected to the top of the BOP and used to lower the BOP into the sea. Hydraulic hoses and electrical cables are secured down the outside of the riser to connect the BOP to the rig. This allows control of the BOP functions and transmits status information back to the rig. The BOP is lowered down to the template, guided by the guide wires. At the top of the riser is a special joint called a *telescopic joint*. This fastens under the rig and it opens and closes to

Drilling Technology in Nontechnical Language

```
Wellhead housing

Outer profile for the
BOP to latch on to

Landing profile
for the next
casing hanger

Surface casing
```

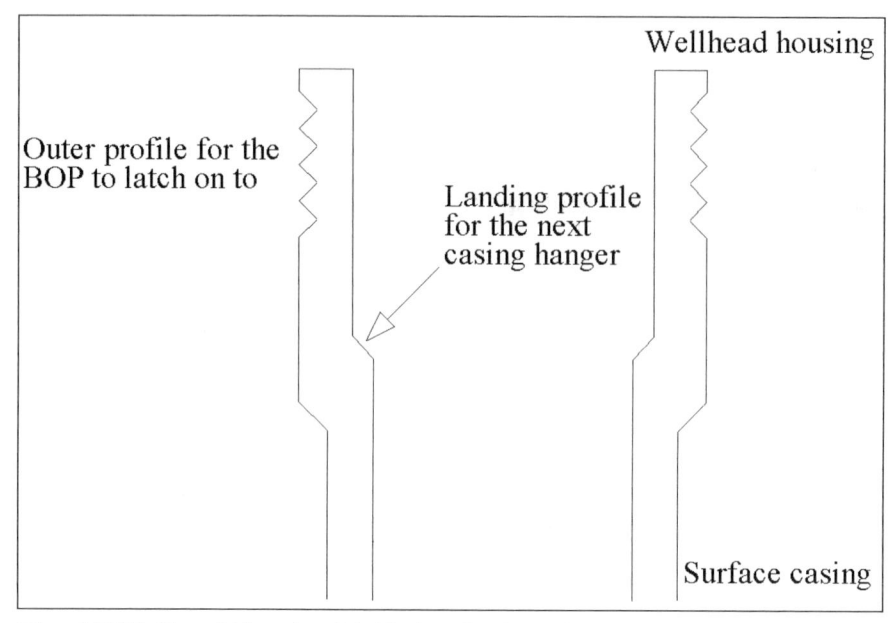

Fig. 4-7 Wellhead Housing Welded to Surface Casing

allow for the up and down movement of the rig once the BOP connects to the wellhead housing. The BOP is lowered so that the hydraulically controlled latch goes over the wellhead outer profile, then the latch is closed and forms a pressure tight seal (Fib. 4-8).

The riser is supported by tensioner wires attached to the bottom part of the telescopic joint. These tensioner wires are attached to pulleys on the ends of hydraulic rams, which are powered by pressurized nitrogen cylinders. In effect, the tensioner acts just like a very powerful spring, the strength of which is set by adjusting the nitrogen cylinder pressure. In very deep water, the tensioner force that is required to support the riser weight is quite high. The riser pipes have shaped "floats" attached to their outside, which reduces the amount of tensioner force needed. As the rig moves up and down ("heaves"), the tensioners allow the wire to respond and maintain the correct force to support the riser (Fig. 4-9).

In summary, the rig has drilled a hole, cemented a conductor in place, drilled a surface hole, and cemented a surface casing in place. The BOP was tested on the rig then run on riser pipe and latched to the wellhead housing on top of the surface casing. The BOP sits on the seabed, controlled by hydraulic hoses from the rig (Fig. 4-10).

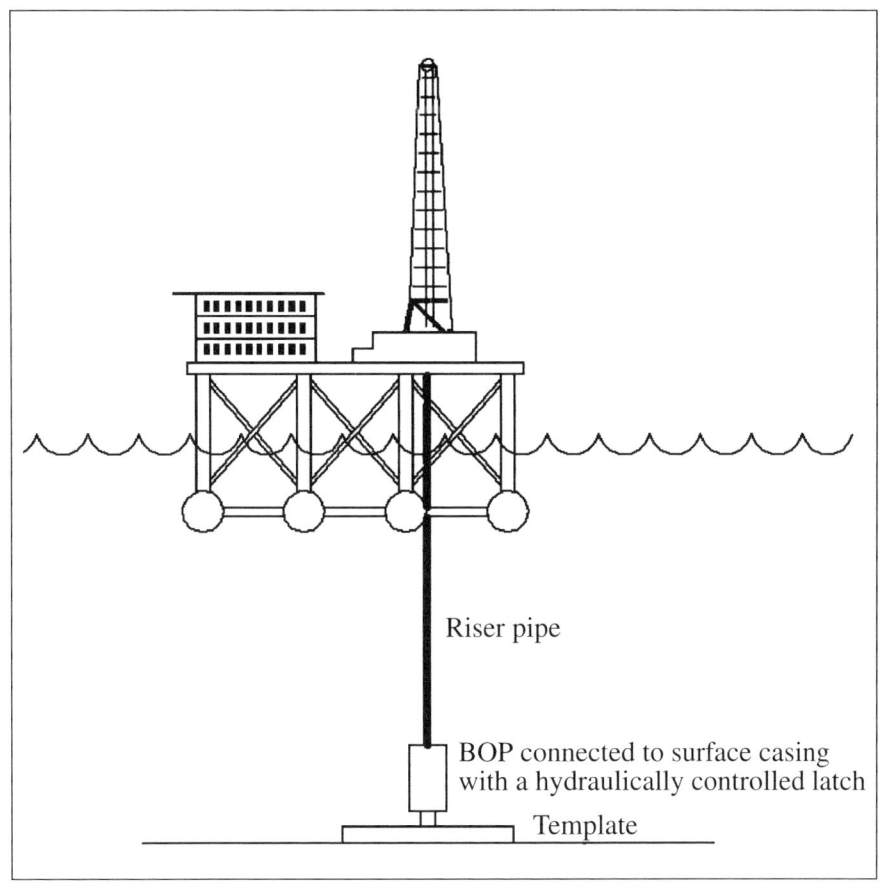

Chapter 4 • Planning and Drilling a Development Well Offshore

Fig. 4-8 BOP Latched Onto Wellhead Housing

Heave compensator

With pipe in the hole, either for drilling or for other operations such as running casing, compensation must be made for rig heave. This ensures that the drill bit stays on bottom with the correct weight on the bit, or that the casing can be landed in a controlled manner. The travelling block (used to suspend and control the drillstring) can be compensated in a similar manner to the way that the riser tensioners work, above.

The compensator cylinder hydraulic fluid is energized by compressed nitrogen. The upward force exerted by the compensator is adjusted by changing the nitrogen pressure. If the upward force just suspends the drill-

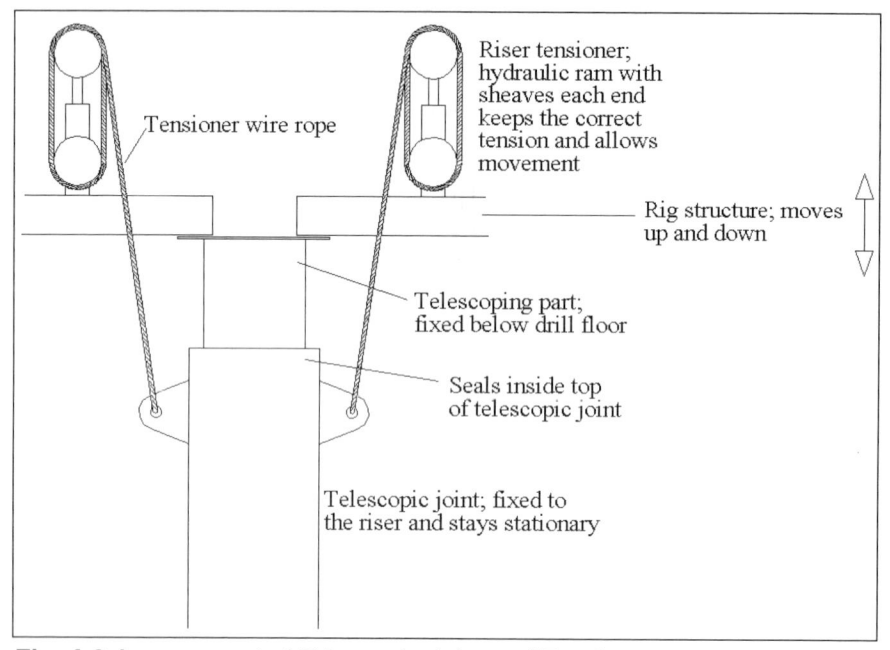

Drilling Technology in Nontechnical Language

Fig. 4-9 Arrangement of Telescopic Joint and Tensioners

string, then reducing the nitrogen pressure to "lose" the desired weight on bit will allow the bit to sit on the bottom with the desired downward force. The amount of heave that can be compensated for is governed by the maximum stroke (movement) of the compensating cylinder rod (Fig. 4-11).

Kicking off the well

When drilling a directional well, the kickoff point has to be carefully chosen. For the kickoff to be controllable, the hole has to be stable (stays equal to the drill bit diameter) over the build up interval. An enlarged hole makes accurate directional control impossible.

The kickoff should be started and finished in one hole section. It is better not to run casing halfway through the kickoff because if the casing is drilled out and a curved wellbore continued, the drillpipe can wear a groove in the casing shoe. Apart from the damage to the casing, the drillpipe can get stuck in this groove when pulling out of the hole. It is better to finish the kickoff and protect it by running casing reasonably quickly afterwards.

Chapter 4 • Planning and Drilling a Development Well Offshore

Fig. 4-10 Photograph of the Telescopic Joint

    If the well is kicked off at an elevated height and the displacement[3] of the bottom of the well is low, the inclination of the well will be low. Below a 15° inclination, it is difficult to control the azimuth of the well. In that case, a deeper kickoff would be preferable because that will give a higher inclination (which is more controllable) to reach the target.

    On the other hand, the lower the well is kicked off, the more the hole has to be drilled to reach the target. Higher inclinations start to cause a variety of problems, such as more severe wellbore instability and it gets harder to circulate cuttings out of the well (especially at inclinations above 55°).

    On this well, it's better to avoid the complications of drilling a kickoff without fluid returns to the rig. The kickoff will be done in the first intermediate hole section.

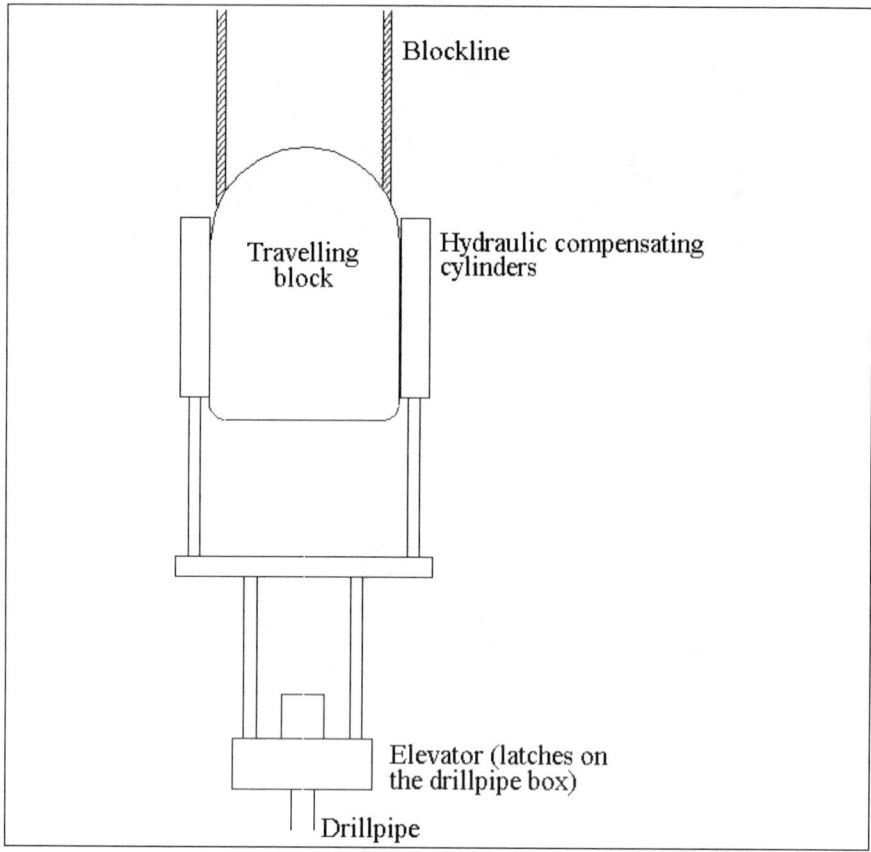

Drilling Technology in Nontechnical Language

Fig. 4-11 Heave Compensator—Principle of Operation

There are quite a few considerations for selecting a kickoff point.

In addition to the kickoff depth, a decision is necessary for how quickly to build up inclination. The severity of a change in wellbore direction is measured in degrees per hundred feet of hole (or per 30m on metric operations). The build up rate should not be too high because this will increase wear on downhole components while drilling. Too low of a build up rate will take more time to complete the kickoff. A build up rate of around 2-1/2° to 3° per hundred feet will work fine. At 3°/100 ft. buildup rate, about 1000 ft. of hole has to be drilled to attain an inclination of 30°.

The surface casing will be drilled out once the BOP is latched and tested. This will be with a normal rotary drilling assembly to the kick off point because the well is vertical above the kick off point. Once the kick off

Chapter 4 • Planning and Drilling a Development Well Offshore

point is reached, the rotary drilling assembly is pulled out of the hole.

From the bit up, the directional assembly will consist of the drill bit, a downhole motor, bent sub, MWD tool (which measures and transmits directional information to the surface), drill collars, and drillpipe. The MWD tool is set up so that the azimuth of the inside of the bend on the bent sub is known. This is important because the drilling assembly will drill in the direction that the inside of the bend points. The inside of the bent sub bend is called the *toolface* and the direction it points at any particular time is the *toolface azimuth* (TFA). The MWD tool transmits three pieces of information to the surface; the wellbore inclination, azimuth at the depth of the tool, and the TFA. This information allows the driller to guide and correct the direction that the well drills.

When drilling with a downhole motor, the motor exerts a torque on the drill bit so that the bit turns and drills. There is also a reactionary torque exerted above the motor. A long string of drillpipe is flexible and will act like a long spring, which will get a little wound up with the reactionary torque from the motor. For this reason, while drilling, the toolface azimuth will move around even though the drillpipe on surface does not rotate. Without a constant surface readout of the TFA, it would be much more difficult to align the drilling assembly in the right direction.

The directional BHA is assembled on the drill floor and the motor is tested by pumping mud through it. If that's okay then the MWD will be added and that, too, will be tested. When all is ready, the assembly is run to just above the bottom of the hole. Once mud is pumped through the drillstring, the motor will start turning and the MWD will start working and transmitting data to the surface. As data is received, the toolface azimuth is known and the drillstring is turned to align the TFA in the correct direction.

With the toolface azimuth aligned, the drillstring is lowered until it touches bottom. The motion compensator is about half open. Then the pressure on the compensator is reduced (so that it supports less of the drillstring weight) the bit takes weight and starts to drill.

While drilling, the main determinant of the build up rate is the angle of the bent sub and the distance from the bent sub to the bit. However, if the weight on bit increases, the build up rate will tend to increase. The adjustment is limited because if too much weight is put on the bit, the torque required to drill exceeds the motor output torque and the motor stops rotating—it stalls.

Drilling the tangent section

After the first build is complete, the well must be drilled in a straight path to the second build position. This is done with a rotary drilling assembly that is designed to drill straight ahead.

In a critical well like this one, where the wellbore has to be drilled accurately, an MWD tool can be run as part of the rotary drilling assembly. The MWD transmits inclination and azimuth information to the surface.

To determine the exact position of an object in space (i.e., an aircraft) it is necessary to know its height above the ground and its geographical coordinates (North-South and East-West). Position within the earth is expressed in the same way; the vertical depth below a particular reference (i.e., mean sea level or the drill floor), called true vertical depth (TVD) and the geographical coordinates.

It is simple to calculate the position of the wellbore within the earth from the inclination and azimuth at known depths. The measured depth (abbreviated to MD) along the hole from the surface to the MWD tool is always known. For each survey there are three items of information—the MD along the wellbore, the inclination, and the azimuth. While there are several ways of calculating the position, the most commonly used is called the "minimum curvature" method because it assumes a perfect arc (segment of a circle) between two survey points (Fig. 4-12).

Knowing the current position, inclination, and azimuth of one survey point, the distance along the hole to the next survey point, inclination, and azimuth of the second survey point, allows for the position of the second point to be calculated. These calculations can be done by hand, but they are somewhat tedious and computer programs are used for this. If two surveys had exactly the same azimuth and inclination then determining the new TVD and coordinates would be simple trigonometry, but even in a tangent section the wellbore is almost never perfectly straight, so perfect circular arcs are still assumed.

In a deviated well, the measured depth is either equal to TVD (in the initial vertical part of the well) or is greater (as the well starts to deviate). Both depth measurements are important; hydrostatic pressures and formation fracture pressures are calculated using TVD, but the lengths of objects placed in the wellbore, such as the casing, are calculated using MD.

Chapter 4 • Planning and Drilling a Development Well Offshore

True Vertical Depth (depth vertically below the wellhead)
vs
Measured Depth (depth measured along the hole)

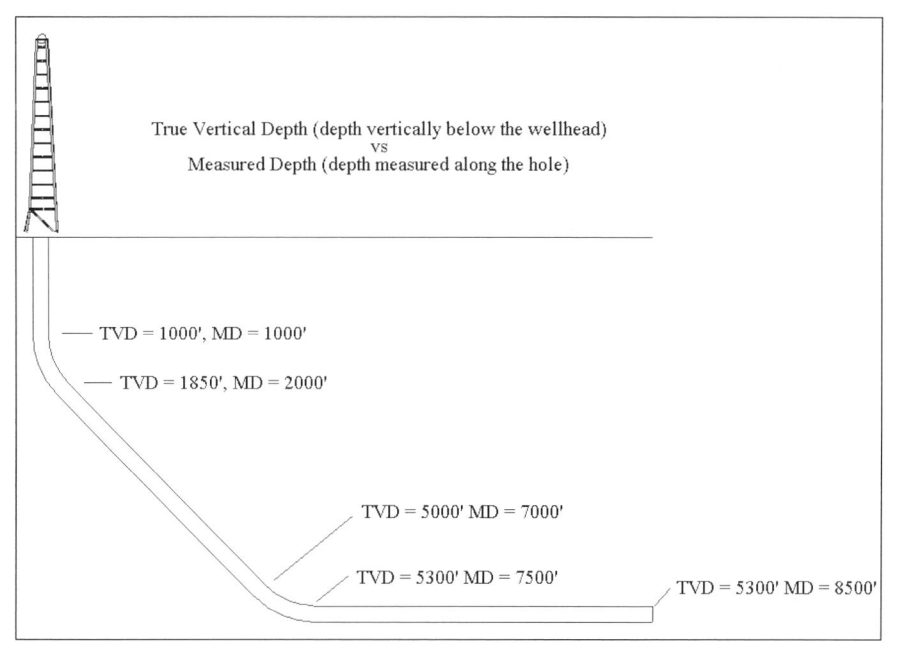

— TVD = 1000', MD = 1000'

— TVD = 1850', MD = 2000'

TVD = 5000' MD = 7000'

TVD = 5300' MD = 7500'

TVD = 5300' MD = 8500'

Fig. 4-12 Calculating Depths in a Deviated Well

Locating the casing point

With the exploration well described in the previous chapter the casing point was planned to be at around a certain depth. The casing was set when a suitable formation was drilled into. In the development well, the formations are reasonably well known and described. The decision as to where the casing will be set can be made in the well design.

As drilling continues, the sequence of formations encountered (identified by rock cuttings sampled at the surface) is compared to the expected sequence and so the *geological* position of the bit is known. If drilling is fast, it may be necessary to stop drilling and circulate "bottoms up[4]" for a sample. Otherwise a formation might be completely drilled through before the cuttings can reach the surface. Often a higher formation can be used as a kind of marker—if the distance from this formation to one of interest is accurately known. There are sometimes more immediate signs—a sudden change in the speed that the bit drills indicates a change of formation drillability[5]. This might identify the next formation in the geological sequence. The change is recognized straight away and may then be confirmed by stopping to circulate bottoms up for a sample of formation.

Once the casing point is reached drilling stops, but circulation continues until the cuttings remaining in the annulus are circulated out and the wellbore is clean. The time to circulate the well clean is greater than the time for bottoms up because the cuttings in the annulus will fall through the rising column of mud so that they rise slower than the mud itself. Circulating until all the cuttings are out is called *circulating clean*.

Logging

On most hole sections of wells (except in development projects where a large amount of wells have already been drilled) logs will be run. Different specialists need information from downhole logs. The geologists want to confirm the properties of the formations penetrated and any fluids within the formations. The drillers want to measure the diameter of the hole using a tool called a *four-arm caliper*, which measures using two pairs of arms so that the hole size is measured in two places. This will allow accurate cement volumes to be calculated. It also indicates how stable the wellbore is and if the wellbore is more stable in one direction than the other.

The drilling engineers also want to improve drilling performance on the next well. It is very useful to measure rock mechanical properties, such as compressive strength. Compressive strength can be calculated by sending a sound wave into a formation and placing a microphone some distance away along the wellbore. The properties of the sound wave are modified by travelling through the rock. This change can indicate, among other things, compressive strength.

Logging a directional hole is more difficult than logging a vertical hole. Any roughness of the borehole wall will impede progress of the logging tool. The tool may stand up on a ledge and not run any further. It is more likely to get stuck as it's own weight presses it into the side of the hole. Generally if the wellbore is smooth, logging tools can be successfully run on wireline up to around a 60° inclination.

In bad wellbore conditions, or if the hole inclination is too high for wireline logging, the logging tools can be attached to drillpipe and run in the hole with the electrical cable inside the drillpipe. Logging on drillpipe takes a long time and so it is very expensive. In this case, only the most important logs needed for decision-making on the well will be run.

An alternative is to run a logging while drilling (LWD) tool as part of the drilling assembly. LWD tools were fairly crude when they were first developed, but modern LWD tools can take logs of such sufficient quality that they can replace wireline logs in some cases.

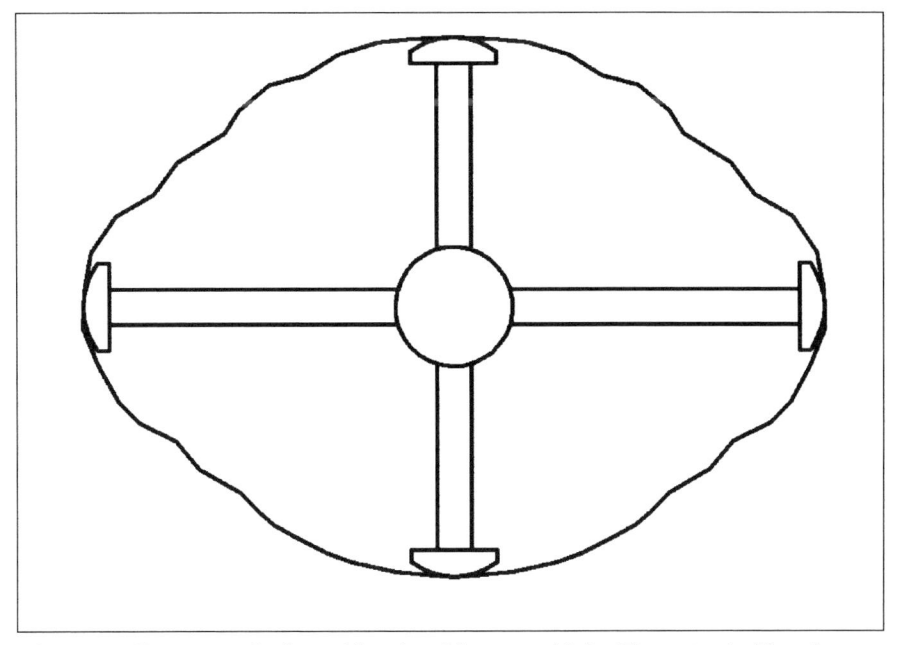

Fig. 4-13 Four-arm Caliper, Used to Measure Hole Diameter in Two Axes

Bringing the well to horizontal

The well is now drilled at an inclination, towards the second kickoff point. The problem with the second kickoff is that it must be exactly right. If the well becomes horizontal just a few feet too low or too high, it will take a lot of drilling to get the well back to the desired TVD. If the formations above the reservoir are just a few feet thicker or thinner, or if their dip angle is just a little bit different to what is expected, the well might end up in the wrong place.

To guarantee accurate knowledge of the downhole geology, the well continues straight past the planned second kickoff point and into the reservoir. Logs are then run and the real situation can be identified. The drilling assembly is pulled out and plain pipe is run in (without a bottom hole assembly). Cement is pumped to the bottom of the well so that the top of the cement is above the required kickoff point and the pipe is pulled out of the hole. Now a steerable motor assembly is run in and the cement is drilled to the point where the kickoff should start. During this time, the drillstring is rotated so that the steerable motor drills straight (Fig. 4-14).

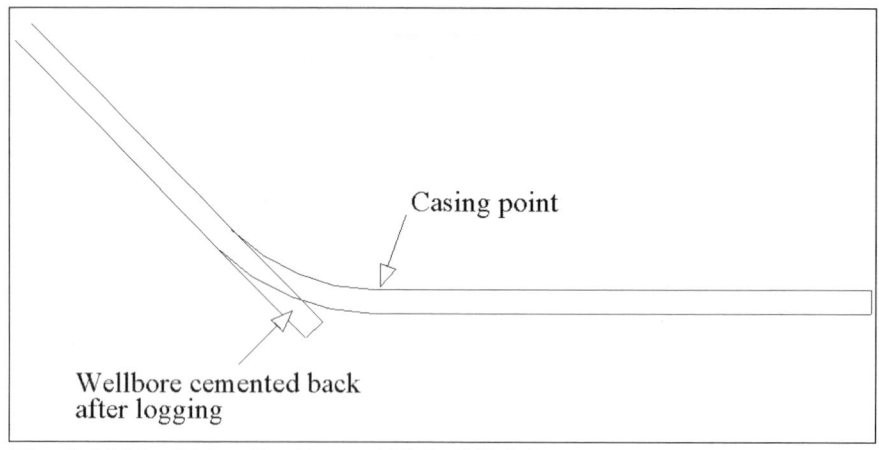

Drilling Technology in Nontechnical Language

Fig. 4-14 Identifying the Second Kickoff Point

It is important that the cement is designed with a higher compressive strength than the formation. If the cement were weaker, it would be difficult or impossible to leave the old wellbore and start a new one.

Once the kickoff point is reached, the steerable motor toolface is aligned in the proper direction and the well starts to deviate away from the original wellbore.

The steerable drilling assembly will include both an MWD tool (to measure directional parameters) and an LWD tool (to confirm the geological position). The well is drilled to become horizontal at the exact TVD (to within four or five feet, possibly less) that is required by the reservoir engineers. At this point, casing can be run into the reservoir to protect the well and to isolate the reservoir from the formations above.

The reason for drilling horizontal wells is to increase the production potential of the well. If a beautiful horizontal well is drilled in the reservoir it's not good to run steel pipe inside it and pump cement around it. Cement has a nasty habit of plugging the reservoir and restricting production. Instead, the well will be left to produce from the well as drilled, if the rock is strong enough to produce without collapsing or producing solids. Otherwise, some kind of screen can be run that is not cemented in place, but that holds back any produced solids and supports the hole if it starts to collapse.

Drilling the horizontal section

Production casing is run and cemented with the casing shoe in the reservoir and horizontal hole. A steerable motor assembly with MWD and LWD is run in to drill the final horizontal hole through the reservoir.

The LWD tool in this well has two sensors—a resistivity sensor and a gamma ray sensor. If the drilling assembly drops towards the oil-water contact, the resistivity reading will decrease. If the drilling assembly builds up towards the shale on top of the reservoir, the gamma ray reading will increase.

Shale minerals are naturally radioactive—they give off gamma rays which can be detected. The level of radioactivity is not enough to harm humans, which is just as well considering that most of the upper earth's crust consists of shales! It is also possible to tell from the gamma radiation spectrum (the frequencies present in the radiation) what type of mineral is responsible for the radiation, so a particular shale can be recognized by its gamma radiation spectrum.

Using our knowledge of the reservoir structure, the LWD and MWD readings allow accurate navigation in the reservoir. It is possible to drill with a high degree of accuracy—within three or four feet.

It's very important to minimize damage to the reservoir as it is drilled through. Normally, a well is drilled so as to remain overbalanced on the pore pressures with the mud hydrostatic pressure, but this pressure will tend to force mud and mud solids into the formation. It is possible to drill the reservoir section underbalanced; that is, the formation pore pressure exceeds the mud hydrostatic. The reservoir will continually flow while the well is drilled. If the problems of hydrocarbon production while drilling can be handled, it would almost guarantee that there will be no significant damage to the formation face. There are horizontal wells drilled in Texas which produce enough oil during drilling the reservoir that the well is paid for by the time the well reaches the target depth!

On this rig, there is no facility to handle produced hydrocarbons, so the option to drill underbalanced is not available. The fluid used for drilling the reservoir will be a special mud with no clay solids added. Clays are especially damaging because they plug pore spaces and cannot be removed with acid. Instead, acid soluble solids such as calcium carbonate can be used. These can later be removed by pumping acid down the well.

Once the well is drilled to the target depth, the drillstring can be pulled out. A sand screen is run in to the bottom of the well, over the full horizontal section, to prevent sand from being produced with the oil. This is fixed in place using a hanger (somewhat similar to a liner hanger).

Suspending the well

The well needs to be left in a safe and stable condition while further wells are drilled and later while the platform is put in place on the template. The practice of leaving a well for later re-entry is called *suspending the well*.

After placing the sand screen in the reservoir, the well has special plugs ("bridge plugs") set in the production casing. Cement will be placed on top of at least one of these plugs, as extra insurance that any gas coming into the well cannot migrate to the surface. These can be drilled out later on.

Most operators and government regulations require at least three physical barriers to prevent hydrocarbon from reaching the surface. In this well, one barrier is provided by the hydrostatic pressure of the fluid in the well; one by the bridge plug and one by the cement plug.

Finally, a *suspension cap* is placed on top of the wellhead. This is a special cover that latches on to the wellhead. Corrosion inhibitor fluid is pumped into the cap to help resist corrosion during the time that the well is suspended.

Chapter Summary

This chapter has examined tools and techniques used to drill a horizontal production well from a floating rig through a subsea template and to suspend it for later re-entry.

The equipment required to allow movement between the rig and the seabed was discussed in detail.

Glossary

[1] **Appraisal.** Once a discovery is made by an exploration well, further wells are drilled to "appraise" the discovery. These wells are called appraisal wells.

[2] **Rotary drilling assembly.** A configuration of drill collars and other downhole tools that drill by rotating the drillstring from the surface, as opposed to an assembly that powers the bit with a downhole motor.

[3] **Displacement.** The horizontal distance between the target and directly below the wellhead.

Chapter 4 • Planning and Drilling a Development Well Offshore

4. **Bottoms up.** An expression referring to the time or volume to circulate in order to bring mud at the bottom of the well up to the surface. If circulating at 600 gallons a minute and the annular volume is 24,000 gallons, it will take 40 minutes to circulate bottoms up.

5. **Drillability.** How easy a formation is to drill. Closely related to the compressive strength of the formation; a lower rock compressive strength gives a higher drillability. There are exceptions—a PDC bit might go from a high strength formation, where it drills fairly fast because the formation is suitable for PDC drilling, to a formation of lower compressive strength, which does not favor PDC drilling. In this case, drilling speed might decrease rather than increase!

Chapter 5

Rig Selection and Rig Equipment

Chapter Overview

This chapter will discuss how a rig might be selected. The different types of rigs will be described and an indication of their comparative costs estimated—though the actual rate for a rig can vary widely depending on the state of the industry. Major items of rig equipment will then be discussed in sufficient detail so that their function within the rig operation and their operating principles can be understood.

Selecting a Suitable Drilling Rig

The major classification of drilling rigs deals with the environment that the rig has to work in. The common classifications of rigs are illustrated in Figure 5-1.

In addition, there are some special units available for working on oil and gas wells which would not be classed as rigs—coiled tubing units (CTU)[1], snubbing units[2], and workover hoists[3].

Also, the maximum load that can be suspended within the derrick or mast should be considered. Deeper wells generally require a higher derrick load rating because the well will require deeper strings of casing to be run, which will weigh more.

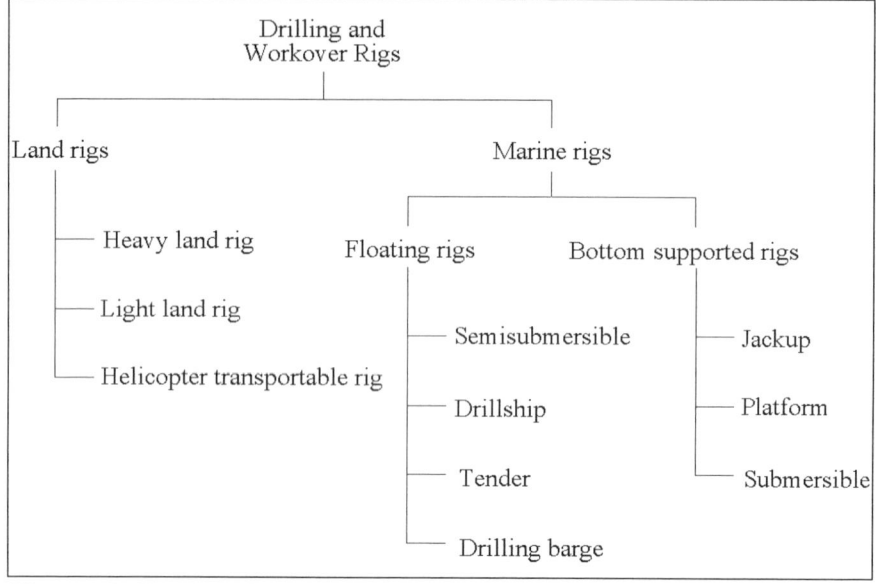

```
                    Drilling and
                    Workover Rigs
                          |
        ┌─────────────────┴─────────────────┐
    Land rigs                            Marine rigs
        |                                    |
                                ┌────────────┴────────────┐
    ├─ Heavy land rig      Floating rigs          Bottom supported rigs
    |                           |                         |
    ├─ Light land rig           ├─ Semisubmersible        ├─ Jackup
    |                           |                         |
    └─ Helicopter transportable rig                       |
                                ├─ Drillship              ├─ Platform
                                |                         |
                                ├─ Tender                 └─ Submersible
                                |
                                └─ Drilling barge
```

Fig. 5-1 Drilling Rig Classification

The cost of the rig is an important consideration. A better, more highly-equipped rig will cost more, but the performance gain should offset at least some, if not all, of the extra overall cost.

The requirement for a rig is dictated by the drilling program, working environment, rig availability, and cost.

Classifications of Drilling Rigs–Description

Each classification of rigs shown in the diagram above is described below.

Heavy land rig

A heavy land rig will be suitable for drilling deep or very deep wells (more than 10,000 ft.). The maximum load that the derrick is capable of suspending will equal or exceed 1,000,000 lbs. The rig will have two, possibly three, high pressure pumps (described later) for circulating the drilling fluid around the well. The BOPs available for use as drilling progresses will be high pressure—rated to 10,000 psi and possibly higher. Storage capacity for consumable fluids and powders will be high. It will need to store and

Chapter 5 • Rig Selection and Rig Equipment

use large volumes of drilling fluid, diesel oil, cement powder, and mud chemicals. The rig systems have to be able to transfer and mix these to produce the required cements and drilling fluids. The rig is transported by a fleet of trucks to the required location.

A heavy land rig might cost around $30,000 a day.

Light land rig

A light land rig will be suitable for drilling shallower wells (less than 10,000' depth). It might also be used for working over[4] existing wells. It will most likely have two high pressure pumps for circulating drilling fluid. Capacities, in general, will be lower than those of a heavy land rig. This rig is also transported by a fleet of trucks to the required location.

Heavy and light land rigs might include accommodation for the crews, or the crews might be accommodated elsewhere while not working (in a hotel or a central camp, for instance) (Fig. 5-2).

A light land rig might cost around $15-$20,000 a day.

Fig. 5-2 Land Rig in the Desert. Photo courtesy of Schlumberger

Helicopter transportable land rig (also called "heli-rig")

In remote areas where suitable roads do not exist, a rig can be placed on location by helicopter. A heli-rig can be broken down into small packages; the maximum package weight will be around 6,000 lbs. Heli-rigs are used in jungles and mountainous regions.

On heli-rig operations, everything is transported by air. The heaviest loads are lifted early in the morning, at first light, because the air is coolest and the performance of the helicopter is the highest in those conditions.

A helicopter will generally be stationed at the wellsite when not in use, so that it is immediately available for medivac[5] in the unfortunate event of an accident.

Heli-rigs will include accommodation for the crews.

A heli-rig might cost around the same as a heavy land rig, depending on the capacity.

Semi-submersible

A semi-submersible rig can be very large. The rig sits on steel columns (between three and eight of them), under which are buoyancy chambers (called *pontoons*). When under transport between locations, the pontoons are empty (or filled with water, as necessary, for stability) so that the rig floats high out of the water.

Once the rig is in position over the wellsite, ballast water is pumped into tanks located within the pontoons and columns so that the rig becomes lower in the water. In this position, the rig will move less than it would in the transport state when influenced by the waves and wind. A large semi-submersible rig can continue to operate in pretty bad weather.

A large semi-submersible rig may be capable of carrying most or all of the equipment and supplies needed for drilling an entire well. If an exploration well were required in a remote area then this would be a serious advantage, as the rig could pick up supplies at a suitable place and proceed to the well location without requiring supply boats being mobilized over large distances.

Accommodation for all personnel is available on the rig. There will also be catering and some leisure facilities, such as a gym, cinema, library, and video room. Many of the crews and supervisors will work on a rotation of 28 days onboard followed by 28 days off.

Semi-submersible rigs may be self-propelled or they may have to be towed by tugs between locations. On location, they might be anchored in place but they can also be dynamically positioned[6] (Fig. 5-3).

The limit of water depth for a semi-submersible rig is dictated by the amount of riser pipe that can be carried or run or by the capability of the anchoring system if the rig is anchored.

A semi-sub rig might cost between $40,000 and $100,000 per day, depending on the age, equipment, and area of operation.

Fig. 5-3 Semi-submersible Rig with Supply Boat. Photo courtesy of Schlumberger

Drillship

A drillship has a ship-shaped hull. Approximately centrally located is a derrick, under which is a large hole through the hull. This is called the *moonpool*.

Drillships vary in size, but the biggest can carry everything needed to drill fairly deep holes in deep water without resupply. They can be moved quickly between locations without tug assistance. Drillships are often positioned dynamically over the well rather than being anchored in place.

Drillships can be expensive to hire. Expect to pay between $50,000 per day for an old one to more than $100,000 per day for a modern DP drillship.

Drilling Technology in Nontechnical Language

Drilling tender

A drilling tender has a ship or barge shaped hull containing the accommodation and all equipment except the derrick, BOP, and ancillary equipment. The tender is moored against a platform and the derrick is installed on the platform deck. Cables and hoses run from the tender to the rig to provide power, drilling fluid, compressed air, communications, control systems, etc. (Fig. 5-4).

A ramp is suspended from the platform, using cables from posts on the platform to the end of the ramp. To transfer personnel between the platform and tender, it is necessary to climb a set of steps on the front of the tender, step (or jump) onto the ramp and then walk up the ramp stairs to the platform. The ramp is called a *widow maker* with typical offshore humor!

In the North Sea, semi-submersible drilling tenders have been used to support a drilling package placed on a platform.

Fig. 5-4 Drilling Tender with Derrick on Platform. Photo courtesy of Schlumberger

Drilling tenders are used for development drilling from established platforms. They might be inexpensive ($30,000 a day) for a barge type ten-

Chapter 5 • Rig Selection and Rig Equipment

der, as shown in Figure 5-4, to $70,000 a day for a modern semi-submersible tender in the North Sea.

Drilling barge

A barge has a large rectangular hull that floats over the well, anchored in position. A derrick is positioned at the end opposite the accommodation, on a cantilever over the hull. They can drill wells up to approximately 20,000' (Fig. 5-5).

Fig. 5-5 Drilling Barge with Derrick Stowed for Moving. Photo courtesy of Schlumberger

Barges are used in shallow waters (up to 50m) that do not have large waves or strong currents. The only drilling barges currently working are on Lake Maracaibo in Venezuela.

Drilling Technology in Nontechnical Language

Jackup

A jackup rig has a floating hull, usually triangular shaped, but sometimes square. At each corner is a large steel leg. The rig is towed to the wellsite with tugs. Once in position, the legs are moved down until they contact the seabed. By jacking the legs further down, the hull raises up out of the water. This forms a temporary platform (Fig. 5-6).

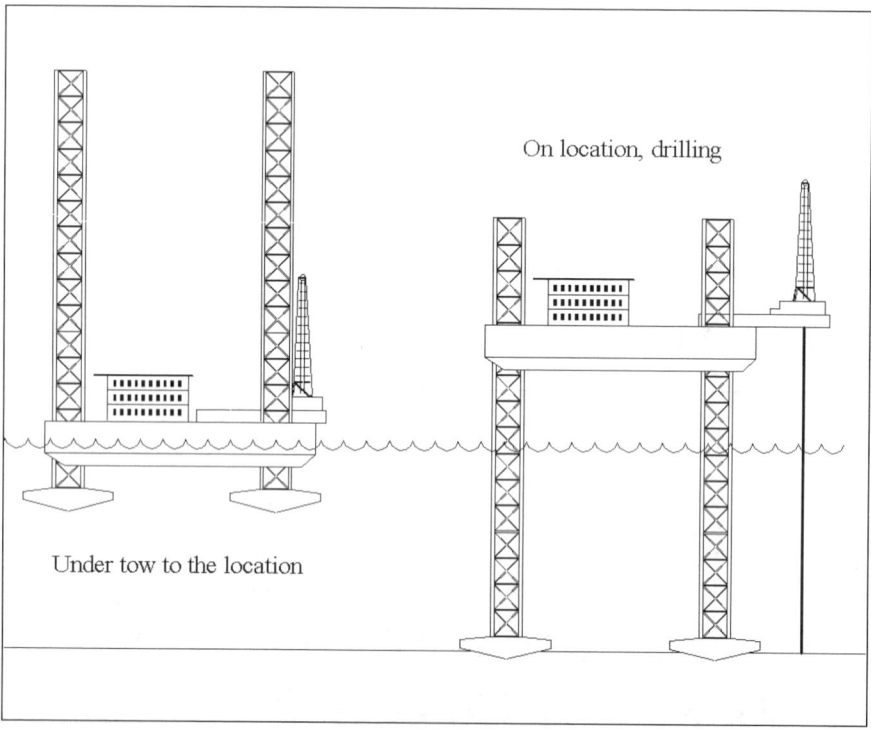

Fig. 5-6 Jackup Rig, Under Tow and on Location

The derrick is located on a large cantilever beam that moves out from the hull, placing the derrick over the side of the hull. This allows a jackup rig to move next to a platform and position the derrick above a well within the platform structure. Exploration wells are drilled by spudding directly into the seabed; a platform is not installed until a decision is made to develop any hydrocarbon discoveries (Fig. 5-7).

A jackup might cost between $30,000 to $70,000 a day, depending on age, capacity and equipment.

Chapter 5 • Rig Selection and Rig Equipment

Fig. 5-7 Jackup Rig Drilling over a Platform. Photo courtesy of Schlumberger

Platform

A platform is fixed in position on the seabed. A conventional platform is built on a framework of legs and bracing tubes. It may include accommodation and a complete self contained rig package, as opposed to a platform where the wells are drilled with a jackup or tender (Fig. 5-8).

A conventional platform could be installed with the capacity to drill quite a lot of wells. A large platform might have the capability to drill more than 30 wells by pile driving conductors into the seabed through guides set into the platform substructure. The rig package sits on top of large steel beams where it is moved around the available well "slots" by using hydraulic cylinders to skid the rig around on the beams. Once the wells are all drilled, the rig package may be removed to reduce platform weight and increase space.

In deep or very deep water, a platform may be designed to float, but is tethered to the seabed with steel pipes. This is called a *tension leg platform* (TLP). Advantages of a TLP include the following:

- The platform is completed in an onshore yard while the wells may be drilled with a semi-submersible rig through a template. Once the platform is ready it is floated to the location, where the legs are attached and the wells connected back to the platform. The time from the decision to develop the field to producing oil can be relatively short, as platform construction and drilling happens at the same time.

Fig. 5-8 Conventional Platform with Self Contained Rig Package. Photo courtesy of Schlumberger

- After the field is abandoned, the platform can be de-tethered, refurbished, and redeployed to another location or to a yard for disposal. The future cost to the industry of removing fixed platforms worldwide is likely to be huge, even if some of them can be toppled on site and left as artificial reefs to encourage fish stocks. The cost of removing a TLP will be much smaller than for a conventional fixed platform.
- A TLP platform could be installed in water as deep as 10,000'. This is far deeper than any fixed structure that could be used.

Submersible

Two types of submersible rigs are used:

- A barge type structure with a flat bottom that is towed to a location in shallow, static water and sunk to sit on the bottom in up to about 20' of water. These rigs are also called *swamp barges*. The derrick can be adjusted to account for slight tilts caused by the bottom being at a slight gradient. These rig types are not common with only 12 operating worldwide.
- A structure similar to a lightly built semi-submersible, with columns and pontoons. The rig is floated to the location where it takes on ballast to sink to the bottom in up to 130' of water. The bottom has to be fairly level. Deck load capacity while moving is low but once on bottom, this type of submersible can take on board a large weight of supplies. In the water depths these rigs would operate in, a jackup type rig would also be able to operate well

Rig Systems and Equipment

Dynamic positioning

With a drillship or semi-submersible rig, it is possible to station the rig above the well by dynamic positioning.

Attached underneath the rig are small propellers mounted on computer-controlled swivelling housings, called *thrusters*. On the seabed, transponders[7] are placed at known locations. The rig positioning system uses these transponders to keep a constantly updated rig position relative to the wellhead. If the rig starts to move away from above the wellhead, the positioning system computers control the thrusters to push the rig back to its defined position.

As described in chapter 4, a floating rig uses a BOP on the seabed and the rig connects to the top of the BOP with a riser pipe. Between the top of the BOP and the riser is a special joint that allows some angular movement between the riser and BOP. However if this angle is exceeded (usually 5°), something will be damaged. If the rig moves outside a certain distance from vertically above the wellhead, a disconnect sequence has to be initiated. This sequence closes the BOP, cuts the drillpipe, disconnects the upper part of the BOP, and disconnects the rig from the well. This disconnect sequence may also be used in an emergency if the rig has to move away from the well in a hurry.

High pressure pumping equipment

While drilling, fluid is pumped down the drillstring and back up the annulus as discussed in chapter 3. Every rig will include among its standard equipment two or more pumps, which together can pump mud at the required flow rates and pressures. Modern pumps are "triplex", that is they have three cylinders and fluid is only pumped when the piston moves forward into the cylinder. These pumps use reciprocating pistons with diameters up to 6" diameter and piston strokes of up to 12", with maximum pump speeds of 120 to 150 strokes per minute (Fig. 5-9).

At the start of the well, the requirement for output pressure is relatively low—less than 1,000 psi. At the same time, the pumps must give a high flow rate, drilling the top hole section of 26" will require flow rates in excess of 1200 gallons a minute. As the well is deepened, the pressure requirement increases but the volume requirement decreases. The pump cylinders and pistons can be changed, so that for higher pressures and lower flow rates, smaller cylinders and pistons can be fitted.

The power required for a given flow rate and pressure can be calculated. The formula for hydraulic horsepower is

Chapter 5 • Rig Selection and Rig Equipment

Fig. 5-9 Triplex Mud Pump Showing the Three Cylinder Blocks. Photo courtesy of Schlumberger

$$\text{Hydraulic Horse Power} = \frac{PQ}{1714}$$

where P is the pressure in PSI and Q is the flow rate in gallons per minute. For a given hydraulic horsepower, pressure is inversely proportional to flow rate—if the pressure doubles, the flow rate must halve for a given HHP. This is fortunate because if a pump is capable of producing a certain HHP at a certain pump speed, the maximum pump pressure can be increased by fitting smaller cylinders and pistons. This also lowers the maximum available flow rate.

The pump output flow is directed into a series of high-pressure pipes leading to the drill floor. On the drill floor is a manifold (called the *standpipe manifold*)—a set of pipes and valves that allow the flow to be directed to different places. Normally, while drilling, the standpipe manifold is set up to direct the flow down the drillstring. Also on the standpipe manifold pressure gauges are positioned that allow the driller to monitor the pump output pressure. This is very important to make sure that drilling continues efficiently and safely (Fig. 5-10).

Fig. 5-10 Standpipe Manifold on a Land Rig

Active mud system

The active mud system is a system of tanks, lines, pumps, and valves that allow circulation of mud around the well and through the *solids-control equipment*. The solids-control equipment is described in the following section.

The active system tanks have some means of allowing mud into the tank and also have a route out of the tank. As discussed earlier, drilling mud usually comprises a liquid (water and/or oil) within which are suspended solids such as clays (for viscosity) and barite[8] (for density control). Other solids materials will enter the mud from the well as drilling progresses. If the mud is allowed to stand in the tanks with little or no movement, solids will start to settle on the bottom of the tank. To avoid this, mud tanks are equipped with agitators. An agitator is simply a large paddle attached to an electric motor—as the motor turns the agitator, the mud is continually

Chapter 5 • Rig Selection and Rig Equipment

moved and this reduces the tendency for solids to settle on the bottom. Usually the corners of a square or rectangular tank will still be relatively static, even with agitators in the tank, and solids will often settle out here.

Settling of solids in the bottom of tanks is detrimental. It reduces the effective volume of the tank. Solids settling will also reduce the density of the mud and this will lead to treatment of the mud to restore the density. Settled solids can also plug pipes and valves in the bottom of the tank. At some point the tank has to be emptied and the sludge on the bottom dug out. It is better if agitation is sufficiently vigorous to reduce settling as much as possible.

It is crucial that the driller knows how much mud is in the active system and in particular is able to identify when the total volume increases or decreases. An increase in mud volume may indicate that fluid is entering the wellbore from a downhole formation (a *kick*); a decrease may mean that mud is being lost downhole. In either case, action must be taken. Each tank in the active system (as opposed to reserve tanks containing mud) must have a means of accurately measuring the volume in that tank in such a way that the driller can be alerted to any change in volume.

The following are two common systems in use:

1. A float is attached to some kind of system that measures its position relative to the bottom of the tank. As the tank dimensions are known, the volume can be calculated from the height from the bottom of the tank to the float (Fig. 5-11).

2. An ultrasonic detector is positioned above the tank. A sound signal is generated and the time for the signal to bounce off the mud surface and return to the detector is measured. Knowing the speed of sound in air, the distance from a known point (the sound generator) to the detector via the mud surface can be calculated and this can be converted to a volume for that particular tank.

The driller has an instrument called the pit volume totalizer (PVT), which can display the volume in any particular tank and can also add up all of the surface active volumes so that any change in the total surface volume may be detected. The PVT also includes alarms that can be set so that changes beyond a set amount lost or gained will cause an alarm to sound.

Drilling Technology in Nontechnical Language

Fig. 5-11 Mud Tank with Float Type Level Indicator. Photo courtesy of Schlumberger

Solids-control equipment

When the drilling fluid leaves the annulus at surface during drilling, it contains rock particles of various sizes that have entered the mud either at the drillbit or after falling off the wall of the hole. If the solids-control equipment does not work efficiently then some of these rock particles can be pumped back down the well and these become ground up even smaller by mechanical action in the well. The smaller they are, the harder it is to remove these particles from the mud when it returns to the surface. Smaller particles also have a greater (bad) effect on the physical properties of the mud which then need expensive chemical treatments to restore the properties to what is necessary to efficiently drill.

The solids-control equipment comprises a series of items to remove progressively smaller rock particles. The first item of equipment that the mud passes through is called a *shale shaker*. The shale shakers are basically a set of vibrating mesh screens that filter out large rock particles and allow the liquid mud to fall through the screens to a tank below. These vibrating screens are designed so that solids filtered out of the mud move along the screen to the edge, where they fall off into a chute and are disposed of.

Sometimes a screen will break, split, or develop a hole and some of the drilled cuttings fall through with the mud into the tank below. It is very important that the drill crews are alert to damaged shale shaker screens because if large amounts of rock bypass the shakers, this will overload the other solids-control equipment and cause problems throughout the system. The screens are replaced if damaged.

Next comes the sand trap. The sand trap is a tank that usually sits underneath the shale shakers. Flow from the shale shakers goes into the sand trap. The purpose of the sand trap is to give temporary protection to the rest of the system if a shale shaker screen splits. In this case, the larger solids will settle in the sand trap and they can be later dumped through a large butterfly valve on the bottom of the sand trap.

Mud from the sand trap flows out at the top, through a cutout in the top rim of the tank. The mud will then flow into another tank. This mud, having routed through the shale shakers and sand trap, will still contain smaller rock particles. This second tank will be used to feed a centrifugal pump that will pump the partially cleaned mud into a hydrocyclone system.

A hydrocyclone is a simple, but ingenious, bit of equipment with no moving parts.

The hydrocyclone comprises a cone with a small hole at the bottom (narrow) end. At the top of the cone is an inlet pipe, positioned so that mud entering the cone swirls around the inside diameter. An outlet pipe exits upward, but the bottom end of this pipe sticks into the top of the cone. See Figure 5-12.

As mud moves around the inside of the cone top, it eventually comes back to the inlet pipe position, where more mud is coming in. This forces the mud stream downwards, into the cone. As the cone gets narrower, the fluid speed has to increase in order to accommodate the flow rate. Very high centrifugal forces are exerted on the fluid stream, so the heavier solids particles will move towards the outside of the fluid stream—moving towards the cone. As the fluid stream nears the bottom, pressure builds up to the point where the fluid changes direction and starts back upwards, spiraling up inside the descending mud that stays close to the cone inside surface. The solids particles, being heavier, cannot change direction so readily and are ejected at the bottom of the cone. The cleaned mud stream exits at the top of the cone, out of the overflow opening.

All of this happens very quickly—it will take something like 1/6 of a second from the mud entering the cone at the inlet to it exiting at the overflow.

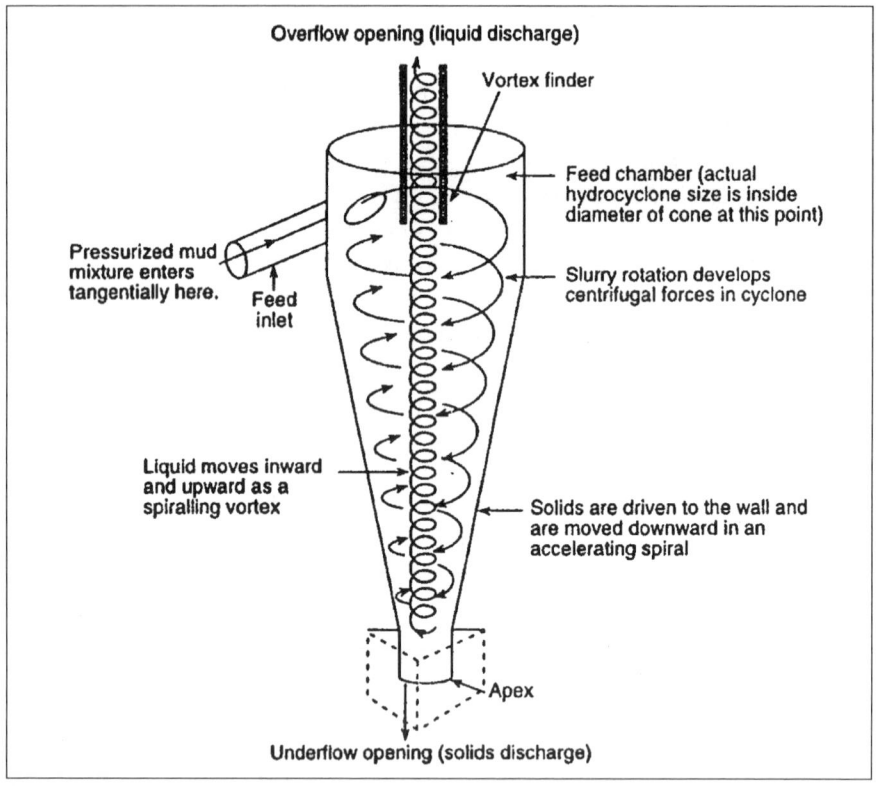

Drilling Technology in Nontechnical Language

Fig. 5-12 Hydrocyclone—Principle of Operation. Picture courtesy of Baroid

Larger cones process larger volumes of mud and remove larger particles. Smaller cones process smaller volumes (per cone), but can remove finer particles. Most rigs will have a set of 3 or 4 large cones, around 12" diameter at the top. These are called *desanders* because they remove particles of sand grain size. Most rigs will also have a set of perhaps 16 or 20 small cones, 4" diameter at the top. These are called *desilters* because they remove particles of silt grain size (Fig. 5-13).

So far, the mud has been processed by the shale shakers, the sandtrap, desanders, and desilters—each stage removing progressively finer particles. If it is felt necessary to remove even smaller particles, a centrifuge (or several centrifuges) may be used.

The cleaned mud ends up in the mud pump suction tank, from where the mud pumps circulate mud around the well again.

Fig. 5-13 A Bank of 4" Desilter Hydrocyclones

Hoisting equipment

The most visible part of a drilling rig is the derrick—the mast that is about 140' taller than the drill floor. The derrick has a set of sheaves on the top, over which steel rope can pass.

The sheaves at the top of the derrick are called the *crownblock*. These sheaves support loops of steel rope that pass through sheaves on the travelling block (Fig. 5-14).

Refer to Figure 5-15 below. The drawworks-drum is usually powered by one or more DC electric motors. As the drawworks reel in the blockline onto the drum, the travelling block moves up the derrick. The drawworks-drum is also controlled by a braking system. To lower the travelling block, the brake is released and the weight of the travelling block and any load under it moves the block downwards and unreels rope from the drum. The brake is used to control the speed of movement. By this means, the rig can pick up or lower down pipe in the well with a very fine control of movement and force.

At the deadline end of the blockline is a sensator[9] that measures the pull on the wire rope. This is connected to a large dial with hands that indicate the weight of the travelling block and any load held under it. This

Fig. 5-14 Crownblock Sheaves at the Top of the Derrick. Photo courtesy of Schlumberger

instrument—called the *weight indicator*—has to be calibrated for the number of times that the line loops around the crown block. The weight indicator is one of the most important instruments that the driller has in front of him at his controls. Every operation involving control of the travelling block is monitored on the weight indicator, which is why it is so large; it's the largest instrument face on the panel (Fig. 5-16).

Rotary equipment

As well as controlling the up and down movement of the drillstring, it is necessary to be able to rotate the drillstring. There are several ways of achieving this rotary movement.

At the top of the rig substructure and the bottom of the derrick is a platform that is called the *drill floor*. This is where the rig crew works most of the time, while tripping pipe in or out of the hole, while drilling, logging, and for most other operations. Somewhere in the middle of the drill floor, directly underneath the crown and travelling blocks, is a piece of equipment called the *rotary table*.

The rotary table assembly comprises a steel housing with an electric motor, gearbox, and power train underneath. This motor and transmission system is connected to a hollow steel cylinder (the rotary table), which can

Chapter 5 • Rig Selection and Rig Equipment

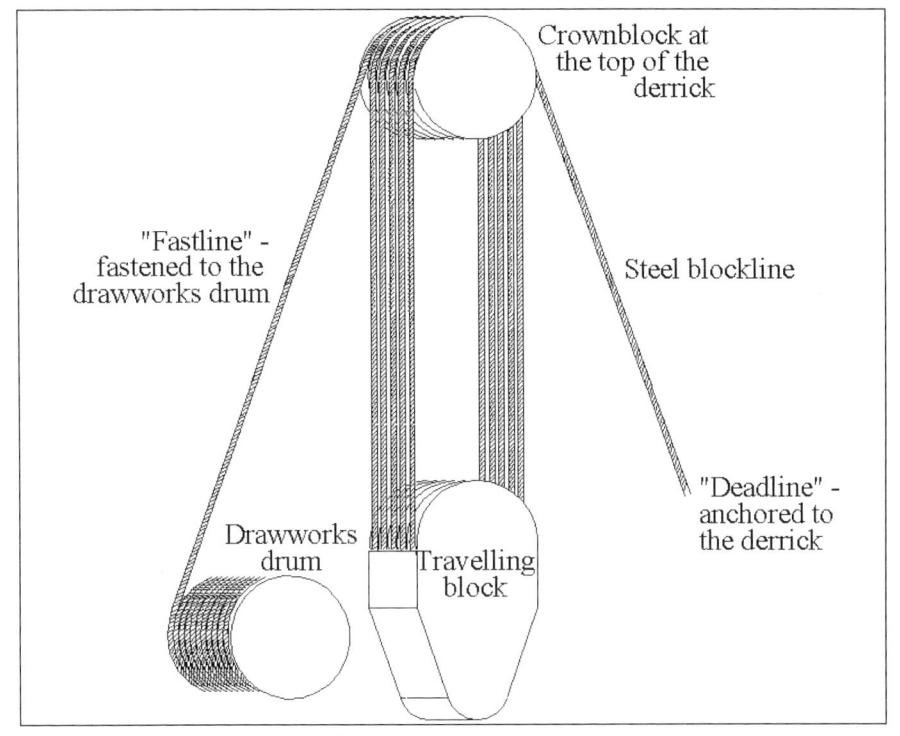

Fig. 5-15 Hoisting System Schematic

then rotate. Bearings underneath support the load of the rotary table and anything supported inside it (Fig. 5-17).

The rotary table has another cylinder, called the *master bushing*, placed inside it and this in turn can accommodate different sizes of sleeves that adjust the size of the hole in the center. These sleeves, called *insert bushings*, have a tapered profile inside (see the Figure 5-17 above). This taper allows the drillstring to be suspended in the rotary table by using a wedge shaped tool called *slips* (described in the section on "Drillpipe Handling Equipment").

On top of the master bushing are four holes spaced 90° apart. These drive holes allow torque to be transmitted to the drillstring by using a special square or hexagonal section of pipe called a *kelly*.

During drilling, the kelly (suspended from the travelling block) is screwed on top of the drillstring. Steel rollers fit on the square or hexagonal faces and these rollers are in a steel cage that has four pins underneath. These drive pins are located in the holes of the rotary table and thus allow

Drilling Technology in Nontechnical Language

Fig. 5-16 Driller at the Rig Floor Controls. Photo courtesy of Schlumberger

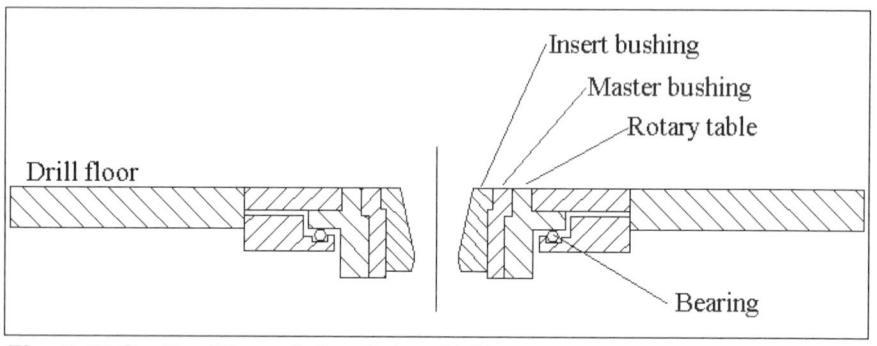

Fig. 5-17 Section Through the Rotary Table

torque to be transmitted from the rotary table to the kelly and then to the drillstring (Fig. 5-18).

With modern rigs, rotary movement and torque is transmitted to the drillstring directly from a motor and transmission system suspended from the travelling block. This is called a *top drive*. It may be electrically or hydraulically powered. If a rig is fitted with a top drive, it will still have a rotary table because sometimes rotation is needed from the drill floor to manipulate tools directly when the top drive could not be used. Also if the top drive should fail, the rotary table and kelly can be used while the top drive gets fixed.

Fig. 5-18 Drill Floor and Rotary Table with the Drillstring Hanging in Slips. Photo courtesy of Schlumberger

Drillpipe

The drillpipe represents a large investment for the drilling contractor. Drillpipe comes in various sizes—probably the most common size worldwide is a 5" outside diameter pipe that has an inside diameter of 4.276" and comes in joints of about 31' in length. The ends of a joint of drillpipe have welded onto them a piece of thick walled pipe on which threaded connections are machined. Joints of pipe are screwed together using these connections. Two of the basic characteristics of drillpipe that identify the pipe are the OD and the type of connection.

Drillpipe is not only specified by the OD and the connection, but also by the type of steel used to make the pipe. In the oilfield, steel used in making downhole tubulars comes in various "grades", where a particular grade is defined by carbon content of the steel, amounts of impurities, heat treatment, etc. For drillers, one of the most important properties of the steel is the strength—how much force can be applied to the pipe before it fails.

Drilling Technology in Nontechnical Language

There are several grades in common use (though other grades are available)

- E75
- G105
- S135

The number part of the grade refers to the minimum yield strength of the steel, in thousands of pounds per square inch. The minimum yield strength is arrived at by testing a sample of the steel in tension and measuring at what force the material increases in length by a set percentage. For "normal strength" steel, E75, the minimum yield is found when the sample has increased in length by 0.5% and the minimum yield stress is 75,000 psi. For higher strength steels it is more—Grade G105 is 0.6% and Grade S135 is 0.7% stretch to give 105,000 psi and 135,000 psi respectively.

If the minimum yield strength of the material and the cross sectional area of the pipe are known, the strength can be calculated. For instance, for 5" Grade E drillpipe, when new, the cross sectional area at an inside diameter of 4.276" is 5.274 square inches. If this is multiplied by the minimum yield of 75,000 lbs., the maximum force that can be applied to the drillpipe is 395,595 lbs. Higher strength pipe has a correspondingly higher maximum force.

When used for drilling, drillpipe is subjected to wear. As it is rotated, parts of the pipe will touch the wall of the hole and the inside of the casing. As drillpipe wears, the thickness of the pipe wall (and hence the cross sectional area) decrease and so the strength of the drillpipe decreases. To allow for this, drillpipe is given a classification that relates to the degree of wear. New pipe is just that—it's within the original manufacturer's tolerances. Next comes premium pipe which has up to 20% uniform wear on the thickness at the OD. Two other classifications are Class II and Class III, but these are rarely used except in drilling very shallow wells or water wells.

Apart from transmitting torque to the drilling assembly and physically supporting the weight of the entire string of pipe, the drillpipe also has to withstand very high pressures from the inside. While drilling deep wells, the pressure at the surface could exceed 3,000 psi, which the drillstring has to be able to withstand.

To specify a particular pipe to use, a drilling engineer has to state the size (OD), grade (of steel), connection type, and classification.

Drillpipe handling equipment

There are three particular items of equipment that are used on a rig to work with drillpipe—the slips, elevators, and tongs.

The *slips* are wedge-shaped pieces of steel that fit inside the insert bushing. The angle of the outside matches the inner profile of the bushings. The slips have steel teeth on the inside face that grip the drillpipe. As more weight is applied to the teeth of the slips, the more the slips are forced down the tapered profile, and the tighter they grip the drillpipe. In this way, the drill crew can easily hang the entire weight of the drillstring (hundreds of thousands of pounds in a deeper well) in the rotary table, using the slips. This is necessary when tripping into or out of the hole to allow connections to be screwed together or unscrewed (Fig. 5-19).

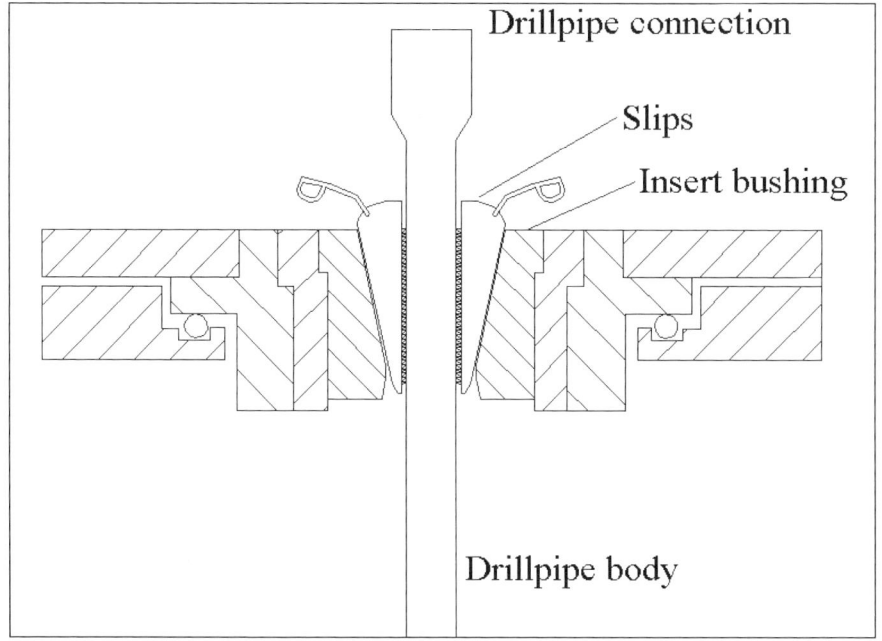

Fig. 5-19 Drillpipe Suspended in the Rotary Table Using Slips

Different designs of slips are used to suspend other tubulars in the rotary table. When running casing or screwing drill collars together, different sizes and types of slips are used. It may also be necessary to remove the insert bushings for the drillpipe and place bushings of a larger ID into the master bushing to accommodate larger sized pipe.

As can be seen from Figure 5-19 above, a drillpipe connection has a larger OD than the pipe body. Underneath the connection is a tapered section, which offers a smooth transition between sizes. On modern drillpipe this taper has an angle of 18°. To lift up the drillstring, a tool called an *elevator* wraps around the drillpipe body and the inside of the elevator has an 18° taper that matches the taper on the drillpipe connection. Elevators are made in two halves with a hinge at one side and a latch at the opposite side. To close the elevator around the drillpipe, it is lowered alongside the pipe (suspended from the travelling block) and the two halves hinged together under the connection taper. The latch closes as the halves come together and locks it shut (Fig. 5-20).

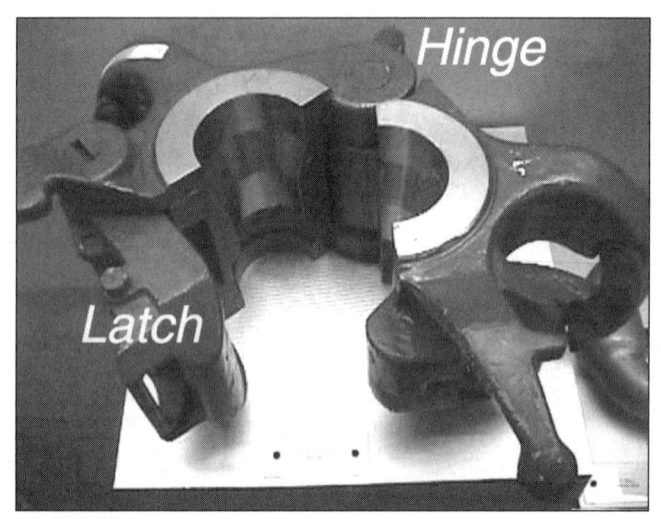

Fig. 5-20 Five-inch Drillpipe Elevator, Open

The slips are used to suspend the drillpipe at the rotary table and the elevators are used to pick up the pipe with the travelling block. The tongs are used to tighten the drillpipe connections.

At the top end of the drillpipe is a threaded female connection, called the *box*. At the bottom end (which normally runs into the well) is a threaded male connection, called the *pin*. When drillpipe is screwed together, the pin is placed inside the box and the upper pipe is turned clockwise to screw the pin into the box. Normally this is done with a pneumatic tool that rotates the pipe. Once the two halves of the connection have come together, it is necessary to tighten them fairly accurately to within a specified

Chapter 5 • Rig Selection and Rig Equipment

torque range. Drillpipe will normally be torqued up to 24,000 ft lbs; large drill collar torques may be in the range of 65,000 ft lbs. to 75,000 ft lbs.

Two tongs are used, one is placed on the box's OD and the other on the pin. The box end *tong* (called the *backup* tong because it backs up the box while the pin moves) is fastened with a length of wire rope to a strong point on the rig floor. The pin end tong (called the *makeup* tong because it's the one that makes up the connection) has a tension sensator attached to it, which measures the pull on a chain that the driller uses to exert a large force on the end of the tong. This chain, the *makeup chain*, is powered by the same electric motor that powers the drawworks.

Torque is expressed as a force multiplied by the length of a lever or arm, hence the term *foot-pounds*. A normal rig tong has a 4' arm, and requires a pull of 6,000 pounds to exert a torque of 24,000 ft lbs. (Fig. 5-21).

Chapter Summary

This chapter discussed rig selection for particular wells and described some of the different types of rigs available. Major items of rig equipment and their operating principles were covered in sufficient detail for the operating principles to be understood.

Glossary

[1] **Coiled tubing unit.** A large drum contains a reel of steel pipe that is quite flexible. This pipe can be lowered into the well to perform various operations, such as pumping fluids, manipulating downhole tools, running logs in difficult or high angle wells, and even drilling

[2] **Snubbing unit.** Allows pipe to be forced into the wellhead under pressure so that jobs can be carried out without having to kill the well and risk damaging it.

[3] **Workover hoist.** Like a very small rig, may be used to pull or run pump rods out or in, or perform small jobs that do not require much power.

[4] **Workover, working over.** If an existing well requires major repair or replacement of the completion, then the old completion is removed from

Drilling Technology in Nontechnical Language

Fig. 5-21 Drill Crew Torquing up a Drillpipe Connection. Photo courtesy of Schlumberger

the well and a new completion is run. This is called "working over" the well.

5. **Medivac.** Short for "medical evacuation". An emergency where a patient requires urgent medical attention that cannot be provided on the rig.

6. **Dynamically positioned.** A floating rig uses thrusters (propellers mounted on swivelling housings) that are controlled by computer. The thrusters keep the rig on station over the wellhead, resisting the push of wind and current. The rig will normally "weathervane"; point into the current to minimize drag due to currents.

7. **Transponder.** A transponder is an electronic device that waits for a coded signal from an interrogating transmitter. When that signal is received, a

reply is transmitted. The interrogating transmitter measures the time for the signal reply and, knowing the speed of the transmission wave, can calculate the distance from itself to the transponder. If four transponders are placed on the seabed a reasonable distance apart, the 3D position of the interrogator in relation to the transponders can be calculated.

[8] **Barite.** A naturally occurring mineral of high specific gravity (4.2), chemically barium sulfate ($BaSO_4$). It is used as an additive in drilling fluids and cements to make them denser.

[9] **Sensator.** A tool that measures some kind of physical force and transmits this information to an instrument or control system.

Chapter 6

Drill Bits

Chapter Overview

This chapter will describe the basic classifications of drill bits and the major design features of each type. The process of bit selection, which is actually quite complicated if done properly, will be outlined in sufficient detail to show the main considerations involved. It will hopefully give an accurate impression of the complexity of bit selection and how critical this is to operational economics.

Drill bits can be separated into two major categories—roller cone bits and fixed cutter bits (Fig. 6-1).

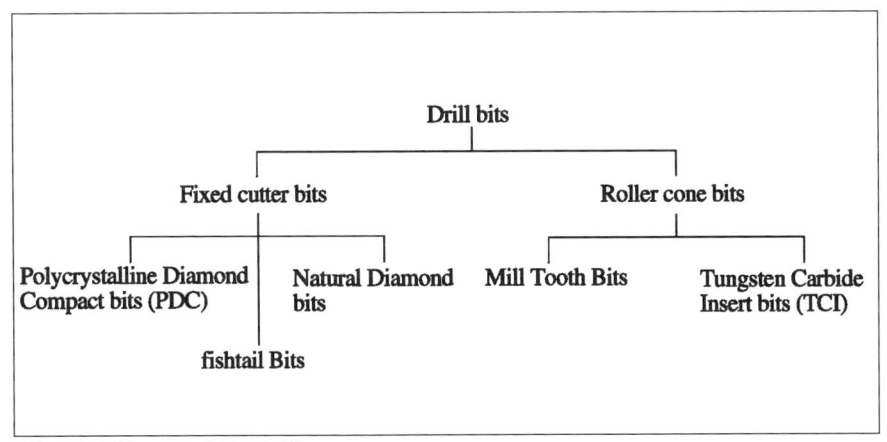

```
                        Drill bits
              ┌─────────────┴─────────────┐
        Fixed cutter bits            Roller cone bits
        ┌───────┴───────┐           ┌──────┴──────┐
Polycrystalline Diamond  Natural Diamond  Mill Tooth Bits  Tungsten Carbide
Compact bits (PDC)       bits                              Insert bits (TCI)
              │
        fishtail Bits
```

Fig. 6-1 Types of Drill Bits

Roller Cone Bits

Roller cone bits have one, two, or three cones that have teeth sticking out of them. These cones roll across the bottom of the hole and the teeth press against the formation with enough pressure to exceed the compressive strength of the rock. Roller cone bits can handle rougher drilling conditions than modern fixed cutter bits, and they are also a lot less expensive. On a relatively cheap drilling operation (land rig) it's almost always more economic to use roller cone bits, except in the smaller hole sizes (8-1/2" and below). Roller cone bits drill faster if the overbalance of mud hydrostatic over formation pore pressure is less. This gives the rig crew an indication of how the pore pressure changes with depth, so an increase in the rate of penetration can warn of an impending kick (influx of fluid from a downhole formation). Fixed cutter bits, by comparison, are less influenced by overbalance and not as useful at indicating possible increases in the pore pressure gradient.

Roller cone bits are available with steel teeth milled from the same block of metal as the cones. These are called *steel tooth bits* or *mill tooth bits*. The other type of roller cone bit consists of steel cones fitted with teeth made of tungsten carbide, which are fitted into holes drilled into the cone surface. Mill tooth bits are very robust and will tolerate severe drilling conditions, but wear out relatively quickly. Tungsten carbide tooth bits will not tolerate shock loadings, but can drill for long distances before wearing out. Of the two types, tungsten carbide teeth bits are more expensive than the same size steel tooth bit.

The outside cutters on a roller cone bit cut at the outside diameter of the hole. These cutters, called *gauge cutters*, are especially vulnerable to wear, and if drilling in an abrasive sandstone these outer teeth lose material and cause the hole to be drilled undergauge[1].

Recently the technology was developed that allows tungsten carbide teeth to be coated with a layer of diamond. These bits are called "tungsten carbide insert" (TCI) bits. When the gauge cutters are coated with diamond it significantly increases the useful life of a TCI bit in abrasive rock (Fig. 6-2).

Chapter 6 • Drill Bits

Fig. 6-2 PDC Bit, TCI Bit, Natural Diamond Bit, and Steel Tooth Bit (Clockwise from top left)

Fixed Cutter Bits

Fixed cutter bits can be divided into diamond bits and polycrystalline diamond compact (PDC) bits. Fixed cutter bits have no moving parts and they can drill for a long time; there are no bearings to wear out, only the cutting surfaces.

Diamond bits drill by wearing out the rock under the bit, producing very small cuttings called *rock flour*. Although diamond bits can drill the hardest rock, they drill slowly and are very expensive. Diamond bits are generally used in the formations with the highest compressive strength, or in formations that are very abrasive, which would destroy another type of bit before it makes much progress. The diamonds used in these bits are naturally occurring industrial-grade diamonds.

Drilling Technology in Nontechnical Language

PDC bits drill with a diamond disk mounted on a tungsten carbide stud (Fig. 6-3). The cutting action is similar to a lathe tool cutting steel. In the right conditions they can drill very fast (more than 100 feet an hour) for great distances (thousands of feet). They are quite costly (especially large ones).

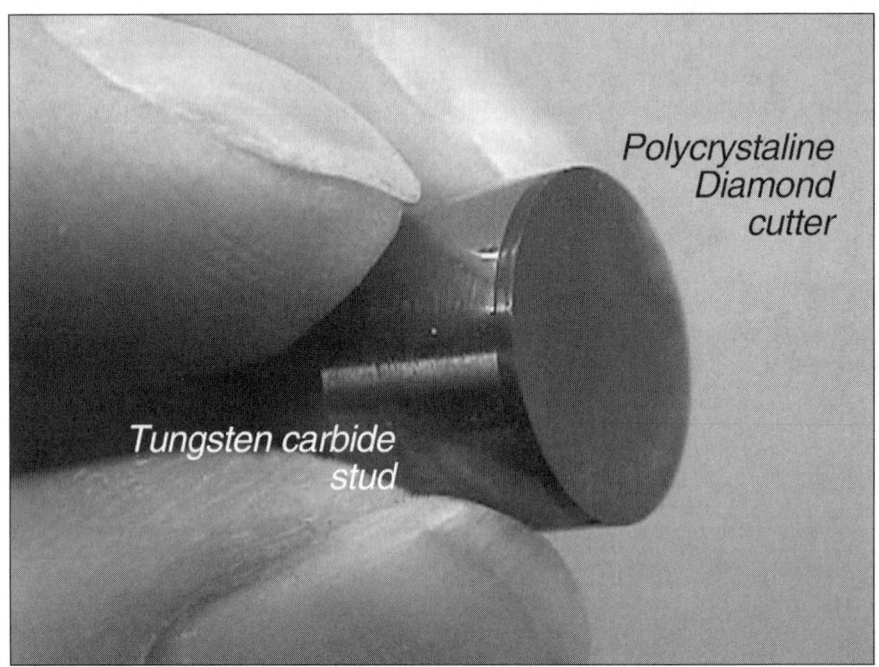

Fig. 6-3 PDC Cutter Mounted on Tungsten Carbide Stud

PDC bits may be constructed from a machined steel body, where the tungsten carbide studs are mounted on steel pegs that fit into holes machined in the body. They may also be constructed from molded tungsten carbide—these are called *matrix body* bits. As always, there are tradeoffs—steel bodies are cheaper to produce than matrix bodies, which are also harder wearing and can be produced in complex shapes more easily.

PDC and diamond bits are made in many different shapes. The shape of a bit will influence whether the bit can be easily made to drill directionally or whether it will tend to drill straight ahead. The shape also affects how many cutters can be mounted on the bit (due to the different surface area). Examples of two extremes are shown in Figure 6-4. The bit to the left

Chapter 6 • Drill Bits

has cutters mounted on the side, combined with the slightly concave, almost flat profile will make this bit easily cut sideways. The parabolic profile bit on the right will be much more directionally stable.

Fig. 6-4 Bit Profiles (Diamond Bits)

One other type of fixed cutter bit should be mentioned. In the early days of the oil well drilling industry, the drill bit was made and sharpened at the wellsite by a blacksmith. These bits resembled a fish's tail when viewed from the side and were called *fishtail bits*. They work by scraping the rock and were only suitable for soft formations. It was not until the advent of the roller cone drill bit (invented by Howard Hughes) that the capability of drill bits extended to drilling greater depths and harder rock (Fig. 6-5).

Core Bits

Core bits cut a doughnut-shaped hole, leaving a column of rock sticking up in the middle of the bit. Behind the bit is a special tube that holds this core of rock and recovers it to surface (Fig. 6-6).

Core bits evolved from a ring structure with rotary cutters mounted on it, through natural diamond core bits to PDC core bits. Most core bits used now are PDC but harder formations still require diamonds.

Core bits often drill faster in the same formation than equivalent regular bit designs. This may be because they have less rock to cut.

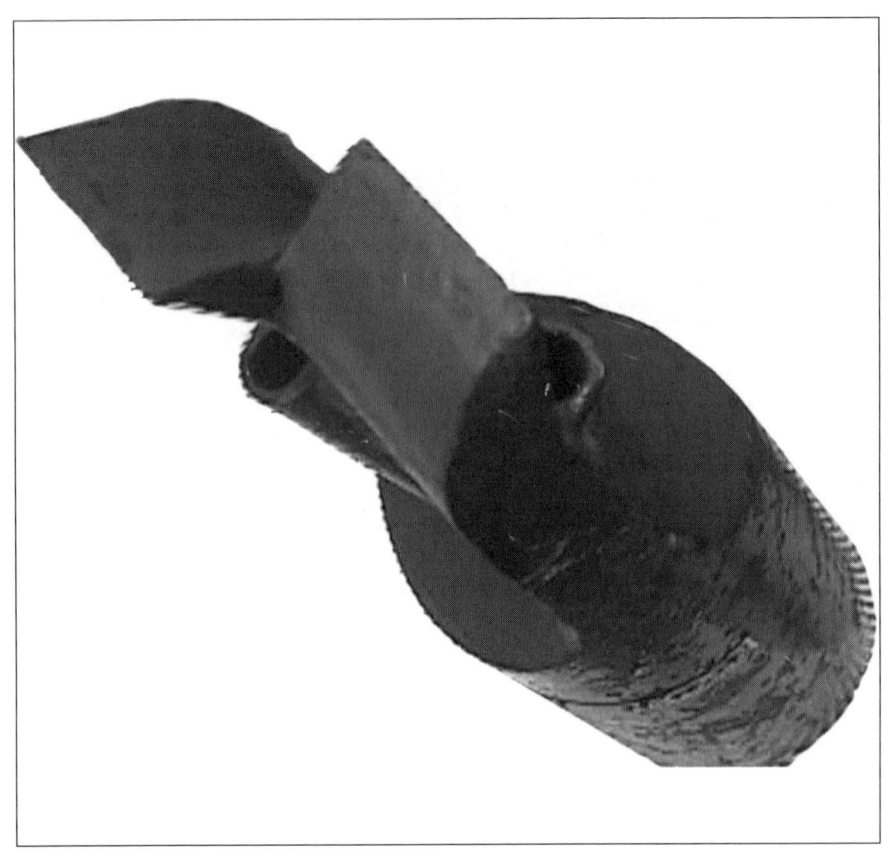

Fig. 6-5 Fishtail Bit

Optimizing Drilling Parameters

As the bit drills, it starts to wear out. The teeth that crush the formation in order to drill will wear. The bearings that allow the cutting cones to turn will wear. The more weight put on the bit and the faster it turns–the greater the rate of wear, yet the faster it drills. As the bit wears out, it will not drill as efficiently, and the rate of penetration[2] (ROP) will decrease. Eventually the bit slows down so much that it is not economic to leave it drilling any longer, so it is pulled out of the hole and changed out for another one.

The faster a well can be drilled the cheaper it is, because time related costs on a drilling operation are high (up to $150,000/day on a modern

Chapter 6 • Drill Bits

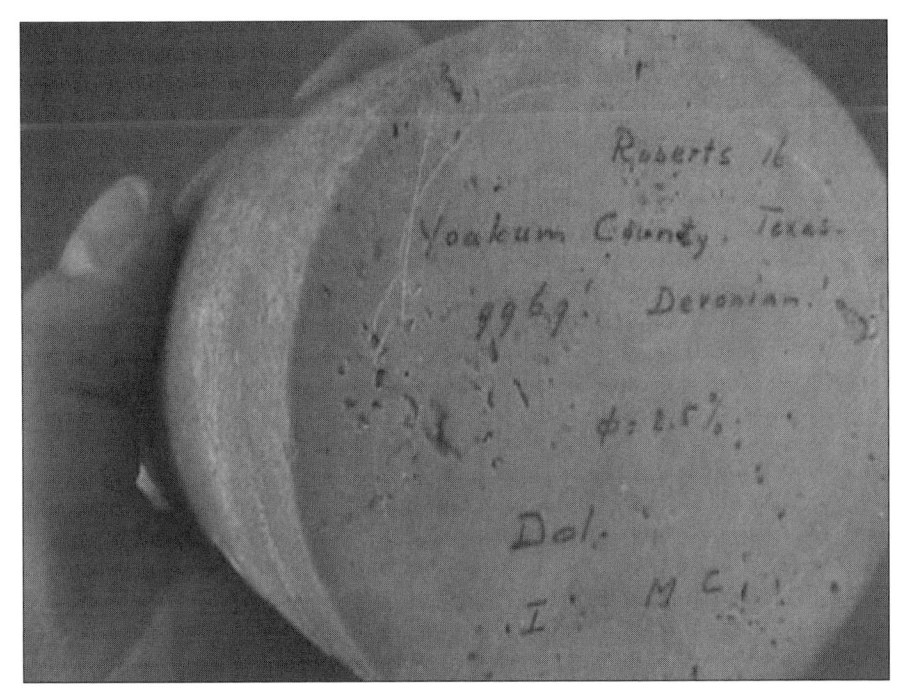

Fig. 6-6 Core Sample

generation offshore floating rig, and a land operation costs about $30,000 /day). To drill faster, in general, the weight on bit (WOB) and the rotary speed [(revolutions per minute (RPM)] can be increased. Increasing either or both of these parameters also increases the rate of wear. So it is very important that the driller finds the optimum set of drilling parameters, giving a good ROP and a moderate rate of wear.

If the WOB is increased, the teeth penetrate deeper into the formation and create larger cuttings. However, there is a point beyond which increasing the WOB does not increase the ROP. This can be because the teeth fully penetrate the formation and the formation then touches parts of the bit that do not drill. It could also be that at the RPM used, the teeth do not have time to penetrate further before they get pulled out again as the bit rotates. So the driller can keep a constant RPM and increase the WOB a bit at a time, each time measuring the ROP. Eventually he can recognize the point at which increasing WOB does not significantly improve performance.

Drilling Technology in Nontechnical Language

If the RPM increases, a tooth penetrates the formation more times in a minute. However, there is a point beyond which increasing the RPM does not increase ROP. This is because the teeth do not have time to penetrate as much as the WOB would otherwise allow. So by first establishing the optimum WOB and holding it constant while increasing the RPM, the driller can find the best combination of WOB and RPM to drill with. In general, drilling economics are reduced by drilling as fast as possible (but within the constraints that other factors might place on ROP), and at the same time not exceeding the optimum drilling parameters, so that the bit drills for a reasonable time.

The procedure for optimizing the drilling parameters is called a *drill off test*.

Effect of mud hydrostatic pressure on ROP

Normally, the hydrostatic pressure of the mud is greater than the formation pore pressure. This is deliberate to maintain control of the well. This overpressure gives rise to a condition known as *chip hold down*.

As a tooth penetrates into the formation, a chip of rock is dislodged. Underneath the chip is the native formation pore pressure and above it is mud hydrostatic pressure. This pressure differential tends to hold the chip in place. If this happens and the chip's departure from the bottom is delayed by a small fraction of a second, then the next tooth down may hit the chip instead of virgin rock. The overall effect is to slow the progress of the bit. In practice with tricone drill bits, it can be seen that there is in fact a strong correlation between mud overbalance and ROP.

This phenomenon can be turned to an advantage. If the bit penetrates a region where pore pressures start to increase, the rate of penetration will also increase. As the natural tendency of a bit in a particular formation is to decrease the rate of penetration with depth, an increase (unless explained by a change in lithology) is one signal that a kick might be imminent and it should be investigated.

PDC bits are less affected by chip hold down, and in exploration wells tricone bits are preferable so that changes in the pore pressure regime can be identified earlier.

Effect of mud solids content on ROP

Low solids content mud will give a better rate of penetration in a similar fashion to low mud overbalance. The reason is not clear, but it is speculated that the mud solids may slow down the equalization of pressure under the chip. With a solids-free system, the drilling fluid can more easily penetrate past the chip to reduce the hold down effect.

Drilling hydraulics

Most drill bits incorporate nozzles that direct the flow of drilling fluid to efficiently clean cuttings from the bottom of the hole and from the cutting structure. If cuttings are not quickly moved away from the bottom, the cutting structure may end up cutting on cuttings, which reduces the ability to cut virgin rock. These nozzles fit into holes (called *nozzle pockets*) on the bottom of the drill bit. This allows the drillers to be able to select nozzles with different inside diameters. A smaller diameter nozzle will increase the speed that mud flows through it (for a given flow rate). If the mud flows faster through the nozzle, it will expend more energy at the bottom of the hole, which may give greater drilling penetration.

The rate of penetration of any drill bit is limited by the ability of the mud to clean the bottom of the hole. Up to a point, increasing the flow rate and increasing the speed with which the mud flow hits bottom will increase the rate of penetration. In softer formations, this force can be sufficient to remove rock by hydraulic force, but unfortunately this often causes an over-gauge hole to be drilled as rock is eroded from the gauge area of the bit.

Grading the Dull Bit

When a bit is pulled out of the hole after drilling, it is referred to as being "dull" (as opposed to "sharp"). The dull bit will have various features caused by downhole conditions encountered by the bit. If these features are properly recognized, together with information recorded while drilling, an accurate picture of downhole conditions can be built up. This then allows a better choice of bit to be made for the same depth in the next well to be drilled. These drilling records and dull bit analyses (called *dull bit gradings*) are therefore very important to improving future performance. It is also important to properly grade the bit in order for the next bit in the hole to be properly selected and run.

For instance, tungsten carbide teeth are quite brittle and can break when shock loads are encountered. If a dull TCI bit has a high proportion of broken teeth, then the possible causes may include excessive vibrations of the drillstring while drilling. Other causes could be hitting the bottom of the hole too hard with the bit, excessive drilling parameters (high WOB and high RPM at the same time), or some steel junk in the hole. If tungsten carbide teeth are not properly cooled (due to insufficient flow rate or clogging up of the bit with formation cuttings) they go through many cycles of heating and cooling as the bit rotates. This causes cracking of the teeth (called *heat checking*) and leads to broken teeth. It's very important that the true cause of these broken teeth is established before decisions are made on the next bit to run in the same well and at the same depth in the next well.

Dull bit gradings are recorded using a standard system of letters and numbers. There are eight characteristics that are noted under this system, four relating to the cutting structure, one relating to the bearings, one to the wear of the bit gauge, one to any other dull features and the last showing why the bit was pulled out of the hole. One problem with dull bit grading is that different people will tend to give slightly different gradings to the same bit. This is mostly because some do not properly recognize the important dull features. It's a very good idea for all dull bits to be photographed and the photo kept in the files relating to the well so that later on questions about the grading might be resolved.

Bit Selection

There are literally hundreds of different drill bits to choose from. Good bit selection is vital in order to get the best drilling performance and to reduce the cost of drilling.

The most important source of data for analysis is the drilling records of other wells in the vicinity. The performance of other bits that have drilled through the same formations in other wells shows what particular bit features are important and which ones should be avoided or are not needed. Dull bit gradings are especially important to see the full story. It is also important to be able to analyze the drilling performance (rate of penetration) and the drilling parameters (WOB, RPM and flow rate) for each foot drilled. This takes some time to complete, especially if there are a lot of wells to analyze, but the work will be amply repaid through optimized performance.

Electric log data can also contribute to bit selection. Sonic logs (which measure the speed of sound through the formations) can be interpreted to give rock compressive strengths, which clearly helps in bit selection. Gamma ray logs analyze clay content and may indicate the best size of PDC cutters to use if PDC bits can be economically run.

Chapter 6 • Drill Bits

If a directional well is drilled, then bits that resist a change of wellbore direction should be used in the straight sections. Bits that do give some side cutting action can be run over the hole sections where a change of direction is required.

On exploration wells where pore pressures are poorly known, it is better to avoid PDC bits because it's important to recognize changes in pore pressure while drilling. As PDC bits are less sensitive to pore pressure changes, predictions are better with tricone bits.

In deeper, small diameter holes, PDC bits start to give some significant advantages, as they have no moving parts. Small roller cone bits have small bearings and the bearing condition, usually monitored by watching torque while drilling, cannot be monitored properly due to the high torque from the long hole. The cost difference between small PDC bits and small roller cone bits is also relatively small, certainly when compared to the larger bit diameters.

In the end, bit selection is an economic decision—which bits are most likely to drill to the next casing, logging, or coring point for the least overall cost?

Drill Bit Economics

If a bit is used past the end of it's economic life, the rate of wear accelerates and eventually parts of roller cone bits might drop off in the hole. This is a problem that will cost a lot of money to solve, because special tools have to be used to recover the bits of junk from the hole before drilling can resume. This is called *fishing*. Fishing can be defined as "a set of activities to remove unwanted material from the wellbore before normal operations may resume". Fishing is discussed in more detail in chapter 13.

How is the economic life of a drill bit measured? By calculating how many dollars are spent to drill the distance with that bit. This calculation is repeated frequently as drilling continues. Within the bit's economic life, the "cost per foot" (or "cost per meter") decreases. Eventually the cost per foot starts to increase. This indicates the end of the economic life of that drill bit. The cost per foot is calculated by adding together the cost of the bit and the cost of the time spent during that bit run (dollars per hourly rig operating cost x hours), this figure is then divided by the distance drilled. The time starts when the new bit is screwed onto the BHA and includes the time taken to run it in the hole. The estimated time to pull out of the hole at the end of the bit run is also added. Here's how the equation looks, assuming the hourly operating cost is $1500,

$$\text{Cost per Foot} = \frac{\text{Bit cost \$} + (\text{tripping hours} \times \$1500) + (\text{drilling hours} \times \$1500)}{\text{Feet drilled}}$$

With a low cost rig (i.e., land rig) and a high cost drill bit (i.e., large PDC bit), the bit would have to drill extremely fast and for a long time if the economics were to compare favorably with a roller cone bit. The dominant factor in this case is the bit cost. However, with a high cost rig (latest generation semi-submersible in a high cost area like the North Sea), a high cost bit is justified if it drills fast and for a long time.

Sometimes the bit is pulled out before the minimum cost per foot is seen—if there are surface indications that the drill bit is damaged, or if the casing point is reached, or if logging is required before the next casing depth. There may be many reasons why the bit run might terminate early. Leaving the bit in after cost per foot starts to increase should seldom be done.

Chapter Summary

All major types of drill bits were discussed in this chapter. The relative advantages and disadvantages of each were covered, demonstrating the many factors that have to be considered when deciding which drill bit to use and what drilling parameters to recommend. Drill off tests, drilling hydraulics, and grading the dull bit were described. Drill bit economics and how to calculate the end of the economic life of a bit were examined.

Glossary

[1] **Undergauge.** Less than the normal bit diameter. If a 12-1/4" drill bit wears at the outer edge and as a result drills a 12" hole, the hole is undergauge because it is less than the unworn (new) bit diameter.

[2] **Rate of Penetration.** The speed at which the bit drills. Measured in feet per hour or meters per hour.

Chapter 7

Drilling Fluids

Chapter Overview

The drilling fluid plays several essential functions in drilling wells. If the mud properties (physical, chemical, and rheological[1]) are incorrect, safety and economics may be severely compromised. The drilling fluid is the single most essential system in safe, efficient, and economic oil well drilling. It is also, in general, the most poorly understood. In some companies the relationship between the quality of the mud and the final return on investment is simply ignored in order to find the cheapest (rather than the most appropriate) mud system, even though the overall cost of the well may be much higher as a result. There are none so blind as those who won't see!

Functions of the Drilling Fluid

The drilling fluid has to carry out all of the following functions at some point in every well drilled:

- Control formation pore pressures to assure proper well control
- Minimize drilling damage to the reservoir
- Stabilize the wellbore so that the hole diameter remains equal to bit diameter, or at least minimizes hole enlargement
- Remove cuttings from under the bit while drilling
- Carry drilled cuttings to the surface while circulating
- Suspend the cuttings to prevent them falling back down the hole when pumping stops

- Release the drilled solids at surface so that clean mud can be returned downhole
- Keep the bit cool
- Provide lubrication to the bit and drill string
- Allow circulation and pipe movement without causing formations to fracture
- Absorb contaminants from downhole formations and handle the difference between surface and downhole temperatures, all without causing serious degradation of mud properties

To perform all these functions requires characteristics that may sometimes be contradictory. This requires the best overall compromise to balance the various needs.

Basic Mud Classifications

Drilling fluids can be divided into seven major classifications, depending on the continuous[2] phase fluid and the type and condition of the major additive within the continuous phase. They are as follows:

1. Fluids with water as the continuous phase and with clays present dispersed throughout the water ("dispersed water-based mud")
2. Fluids with water as the continuous phase and with clays present inhibited from dispersing throughout the water ("non-dispersed water-based mud")
3. Clear fluid systems based on water with soluble salts used to control density ("solids free" systems or "brines"). Brines may include acid soluble solids that can be removed from the reservoir face by circulating acid past the reservoir
4. Fluids with oil as the continuous phase and less than 10% water by volume, with any water forming an emulsion of water within the oil ("oil mud")

5. Fluids with oil as the continuous phase and more than 10% water by volume, with the water forming an emulsion of water within the oil ("invert oil emulsion mud")
6. Fluids with air as the continuous phase ("air drilling")
7. Water-based systems incorporating air present in gaseous form within a liquid ("aerated" and "foamed" systems)

Each of these types will be briefly described. Their particular properties and uses will then be covered.

Dispersed mud

In the first part of chapter 1, clay mineral types and clay hydration were discussed. Some clays react strongly with water—they hydrate—and will expand in the presence of water. Water that is allowed to enter the crystal structure can cause the crystal lattice to expand because of changes in electrostatic forces. This expansion is described as dispersion. Water molecules are polar; that is, the water molecule, while being electrically neutral overall, has positively and negatively charged areas in different parts of the molecule. The polar nature of water can be increased by the addition of alkalis such as the monovalent bases, sodium hydroxide, or potassium hydroxide. The more polar the water, the more reactive clays will disperse.

The potential reactivity of a clay formation will depend on the types of clay present and the physical environment. Some clays are more likely to hydrate and expand than others. One such highly reactive clay mineral in the presence of supplied water is montmorillonite. The montmorillonite crystal structure comprises large, flat sheets of alternating octahedral and tetrahedral layers. For this reason it is often described as *mixed layer* clay. Other types of mixed layer clay also occur (Fig. 7-1).

Montmorillonite is added to the mud to give it certain useful properties. Commercially supplied montmorillonite is known as *bentonite*.

When fully dispersed, clay such as montmorillonite will have its clay platelets completely separated and held apart by negative charges on the faces of the platelets. The theoretical surface area of fully dispersed montmorillonite is around 800 m^2/g. It is this huge surface area of dispersed montmorillonite that causes dispersed mud with added bentonite to become viscous. The clay platelets are also able to plug permeable formations, which can reduce loss of mud to some formations.

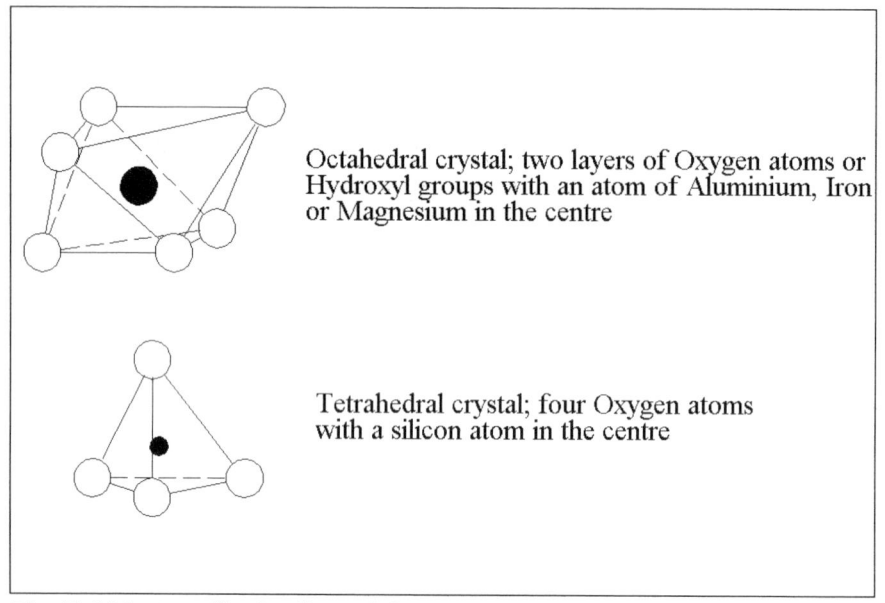

```
                    Octahedral crystal; two layers of Oxygen atoms or
                    Hydroxyl groups with an atom of Aluminium, Iron
                    or Magnesium in the centre




                    Tetrahedral crystal; four Oxygen atoms
                    with a silicon atom in the centre
```

Fig. 7-1 Montmorillonite Crystal Composition

Mud designed so that clays are dispersed (either added like bentonite or from drilled formations) is called *dispersed* mud.

Non-dispersed mud

Mud where the hydration and dispersion of drilled clay is minimized is called *non-dispersed*. There are a number of ways to achieve this. The most common is to limit the amount of water that reacts with the clay by encapsulating the clay with a polymer[3] as quickly as possible, to prevent further access of water to the clay.

The electrical charges on the surface of the clay particle attract sites on the polymer chain that have an opposite electrical charge. The result is that the long chain of the polymer can wrap itself around the clay. Very long polymer chains can hold several clay platelets together, which helps the solids-control equipment to remove drilled clays at the surface. Such mud systems are described as *encapsulating polymer muds*.

Originally the polymers used as drilling fluid additives were naturally occurring starches that were easily extracted, such as cornstarch (first documented use in 1937). Other natural polymers were also tried and entered common use. Now, synthetic polymers are often tailored to specific drilling situations. The monomer units can be similar or very different. The size of the polymer molecule can be controlled. Groups of molecules can be attached to the polymer "backbone" to impart important characteristics.

Polymers can perform several different functions, such as

1. Increase the viscosity[4] of the fluid
2. Increase gellation[5] properties
3. Decrease fluid loss[6] into the formation
4. Act as a surfactant, to allow oil and water to mix together in an emulsion

Solids free brines

Brines are used when working within the reservoir so as to minimize damage to the formation. Sometimes a solid-free or a brine system is used for drilling through the reservoir. Later on during completion or workover[7] operations, these systems are again useful to reduce or eliminate damage.

Brines can be formulated as solids-free systems with density gradients up to 1.07 psi/ft. Solids weighting materials and fluid loss additives that are acid soluble can also be added, such as calcium carbonate and iron carbonate. While being able to overbalance formation pressures, properly designed brines do not create formation damage—either by plugging the reservoir with unremovable solids or by causing reactions with formation fluids or solids. Potential interactions of brines in the reservoir include

- Scale from the reaction of a divalent brine with dissolved carbon dioxide, producing an insoluble carbonate (Divalent brines: those containing calcium or zinc salts)
- Precipitation of sodium chloride from the formation water when it is exposed to certain brines
- Precipitation of iron compounds in the formation resulting from interaction with soluble iron in the completion fluid (most common with zinc bromide ($ZnBr_2$))
- Reaction of formation clays with the brine

- Corrosion of casings and tubulars (not such a problem with monovalent brines)

Oil mud and invert oil emulsion mud

An oil mud comprises various solids and additives mixed into an all-oil continuous phase with little or no water (10% or less by volume of liquid).

With invert oil emulsion mud (IOEM), water is present at more than 10% by volume within the continuous oil phase as an emulsion. The water ("brine" would be more accurate because it will contain dissolved salts) forms tiny droplets that are completely surrounded by the oil. The droplets of brine are held in an invert emulsion in the oil phase because they are coated by emulsifiers. Emulsifiers are surfactants that have an organophilic end and a hydrophilic end to their molecule. Each end aligns itself to be in either the oil phase or the water phase.

While the terms "oil mud" and "IOEM" have specific definitions, the terms oil mud, oil base mud, etc. are interchangeable and really define fluids where the continuous phase is oil, and if an emulsion is present, it is as an invert emulsion (water emulsified in oil). The presence of a water phase in oil mud allows for versatility in control of parameters such as rheology.

Many different oils have been used in the past, including crude oil, diesel oil, oils extracted from fish or plants, and synthetic oils. Some of these oils are toxic, carcinogenic, and flammable, (crude and diesel oils) which are undesirable for safety, environmental, and health reasons.

Air as a circulating medium

It is possible to circulate with compressed air. For this to work, the following conditions are necessary:

1. The formations to be drilled must remain stable without hydrostatic mud pressure to support them
2. There must be no danger of a fluid influx into the well (oil or salt water)

The areas of application are hard, dry formations—dry geothermal zones and dry gas production zones. While drilling through a gas-bearing reservoir, the well produces gas while drilling.

Aerated and foamed mud

Aerated mud involves simply injecting standard drilling mud with air, effectively gas-cutting the returns and lightening the fluid column. The main advantages are

1. Maintaining full circulation in loss zones
2. Increasing ROP by reducing chip hold-down (explained in chapter 6)
3. Reducing the incidence of differential sticking[8]
4. Reducing formation damage

Air may be injected at an appropriate rate in proportion to the mud circulation rate. Generally, the technique is limited to a maximum depth of about 2800', as injection pressures become excessive at greater depths.

In foam mud, the liquid is a continuous phase and contains encapsulated air bubbles within it. The percentage of liquid will vary between 2% and 15% by volume. The technique allows a variable bottom hole gradient of between 0.026 psi/ft and 0.312 psi/ft.

The lifting capacity of stable foam is superior to that of drilling mud; cuttings are circulated out of the well more efficiently with foam. It is possible to displace fluids from the hole using foam. Oil and salt water influxes are likely to destroy the foam stability, precluding the use of foam in those conditions.

Designing the Drilling Fluid

In selecting the most suitable type of drilling fluid, many different factors must be considered. Overall, what is required is a mud system that gives the *lowest overall cost* of drilling each hole section, except through the reservoir. The direct cost of the fluid itself (cost per barrel of mud) is but one component of this overall cost. If serious hole problems occur because the mud was not optimized for the formations (to "save money") then, of course, much more money will be spent than would have been saved on the mud bill.

When drilling through the reservoir, the key is to minimize damaging reactions between the mud and the reservoir, which lower the possible production from the well. If a well loses only 10% of it's potential produc-

Drilling Technology in Nontechnical Language

tion rate due to avoidable damage from the mud, then the cost to the operator in lost profit over the full life of the well will be large.

Mud cost must be considered, but only to choose between technically suitable systems. Therefore, what should happen is that for each hole system all technically suitable alternatives should be defined, then the cost of each can be compared for a final choice.

Physical, rheological and chemical characteristics can be defined for each hole section, leading to a list of requirements for the mud system of choice.

Mud physical properties

Density. Primary control of downhole pressures is obtained with mud of such density as to exert a greater hydrostatic pressure on the formation than exists within formation pores.

The *lower* safe limit of mud density is calculated by the density to balance the formation pore pressure, plus a small additional amount as a safety margin.

Some formations require a minimum hydrostatic pressure to keep them stable. When a hole is drilled through a rock in the ground, the stresses in the surrounding rock will tend to push the rock into the hole. If mud hydrostatic pressure is kept high enough, it pushes back against the rock and so supports it. The required density gradient is likely to be something greater than that required to maintain well control.

The *upper* safe limit of mud density will be given by one of several factors

1. Losses or formation breakdown may be induced if the hydrostatic pressure plus circulating pressure losses[9] exceed formation strength
2. A high mud density will give a reduced MAASP. (MAASP was explained in chapter 3, "Planning and Drilling an Exploration Well on Land"); this means that the well is less able to withstand a kick than with a lower mud density
3. Rate of penetration is generally reduced with higher weights due to chip hold-down (as explained in Chapter 6)
4. Differential sticking becomes more likely at higher mud densities

5. Higher mud densities have higher solids content and this will adversely affect mud rheology, possibly calling for more additives to control

6. Some shales contain tiny fractures; when mud or filtrate[10] is forced into the fractures it lubricates the fracture faces and also changes the stress regime in the near-wellbore zone–the wellbore will become unstable and chunks of the shale will fall off into the well

On balance, the correct density within this range of maximum and minimum will normally be closer to the lower limit.

Fluid loss. The fluid loss property of mud indicates how well the mud forms a seal against permeable formations. To test fluid loss, a sample of mud is placed inside a chamber, which has a standard filter at the bottom. The chamber is closed and 100 psi is exerted on the mud sample. Filtrate is squeezed through the filter into a container below and wall cake builds up on the filter. The standard test measures the amount of filtrate collected in 30 minutes and the thickness of the filter cake in 1/32 of an inch or in millimeters. A description of the filter cake might also be made, using descriptions such as hard, soft, tough, rubbery, firm, etc.

High fluid loss mud will build up a thicker, stickier wall cake that is likely to lead to problems such as differential sticking. Ideally the mud should build up a thin, tough, and impermeable cake fairly quickly.

The test for fluid loss is a comparative test. It does not indicate how much filtrate will actually be lost to the formation, or how thick the filter cake might actually become. These things depend on many factors, such as the actual pressure overbalance, the permeability of the downhole formation, effects of mud flow, or pipe movement eroding the filter cake, etc.

Sand content. Sand is normally the most abrasive solid present in the mud and a high sand content will increase wear on pumps, valves, and other equipment. However, all solids in the mud will contribute to mud abrasiveness. Sand content should be kept as low as possible by using the solids-control equipment properly. The sand content is measured by passing a fixed volume of mud through a 200 mesh sieve into a marked container and the sand content is measured off directly from the marks.

Mud rheology

Rheology should be renamed to "rheallycomplicatedology". It is the science of the deformation and flow of matter. When discussing the rheology of fluids in the well (mud, cement, and brines), what is of interest is the relationship between the flow rate and the pressure required to maintain that flow rate (either in the pipe or in the annulus). The relationships between these properties will affect circulating pressures, surge and swab pressures[11], and hole cleaning ability[12]. To fully specify a drilling fluid that must perform specific functions, the required rheology must also be defined.

In order to discuss rheology, it is necessary to define some terms. First imagine two square plates that are one meter by one meter in size—these plates are one meter apart and that the gap between the plates contains fluid (Fig. 7-2).

When discussing rheology, flow rate is not used; the shear rate of the fluid is expressed. Shear is movement sideways. Shear rate is the rate at which a section of fluid moves sideways (the top plate) relative to a parallel section of fluid (the bottom plate). Now referring to the two plates, the shear rate is expressed as "meters per second (speed of movement of one plate relative to another) per meter (of distance apart)". If the shear rate were 1 meter/sec per meter, then the top section of the fluid moves at 1m/sec relative to the bottom section of fluid (referring to the diagram). If the shear rate profile between the plates is uniform, then the fluid halfway between the plates will move at 0.5 m/sec. The shear rate here will still be 1, calculated as

$$\frac{0.5 \text{ meters fluid speed}}{1 \text{ second} \times 0.5 \text{ meters apart}} = 1$$

the units of shear are $\frac{\text{meters}}{\text{second}} \times \frac{\text{meters} = 1}{\text{seconds}^{-1}}$ or reciprocal seconds

The symbol for shear rate is γ (Greek letter lowercase gamma).

Note that as the meters cancel out within the equation, a shear rate in sec^{-1} is the same whether Imperial or metric units are used.

To force a fluid to flow at a certain shear rate takes a certain pressure. In rheology, the force required is expressed as *shear stress*. Stress is a load for a unit of area. The symbol used for shear stress is τ (Greek letter lower

Chapter 7 • Drilling Fluids

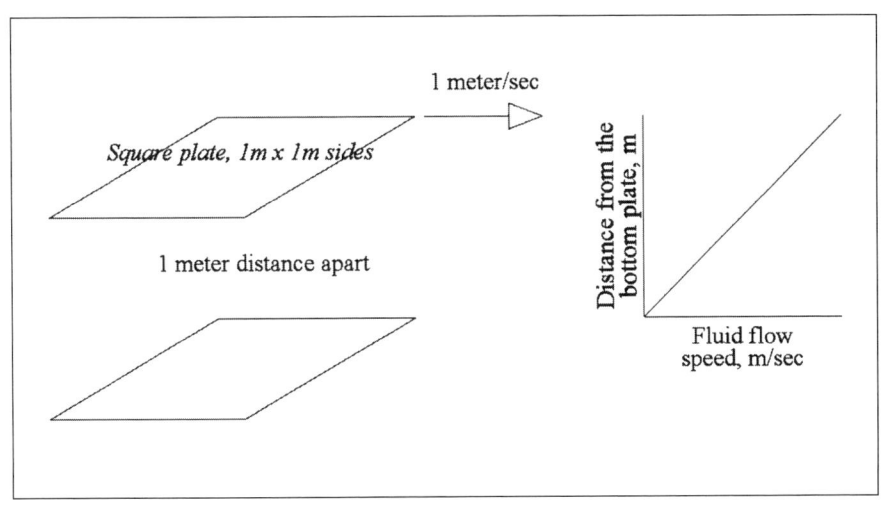

Fig. 7-2 Rheology—Shear Rate

case Tau). The force is measured in pounds or dynes and the area in hundreds of square feet or cm² (Imperial or metric measurements), so stress is expressed as

$$\frac{lbf}{100\ ft^2} \quad or \quad \frac{Dynes}{cm^2}$$

1 dyne/cm² = 10 pascals = 20.9 lbf/100ft² = 0.0000145 psi

Viscosity is the relationship between shear stress and shear rate. It is a measurement of the fluid's internal resistance to flow. The symbol used for viscosity is μ (Greek letter lowercase upsilon).

$$\text{Viscosity } \mu = \frac{\text{Shear stress } \tau}{\text{Shear rate } \gamma}$$

Understanding shear stress, shear rate and viscosity is fundamental to understanding rheology, so read through it again if it is not clear before continuing further.

For completeness, the units of viscosity are now discussed. However, it's not vital that this is understood to discuss rheology.

If viscosity = shear rate/shear stress, then the units of viscosity should be $^{lbf \cdot s}/_{100ft^2}$ (Imperial) or $^{dyne \cdot s}/_{cm^2}$ (metric) ["s" is the symbol for seconds].

Using the conversion figures above, to convert $^{lbf.s}/_{100ft^2}$ to $^{dyne.s}/_{cm^2}$, divide by 20.9.

If pascals were used instead of $^{dyne.s}/_{cm^2}$ (both are metric units of pressure) then metric viscosity could be expressed as Pa.s. To convert $^{lbf.s}/_{100ft^2}$ to Pa.s, divide by 2.09.

Most of the time, viscosity is expressed in the derived unit "centipoise".

$$1 \text{ Centipoise} = 0.1 \text{Pa.S} = \frac{0.209 \text{ lbs}}{100ft^2} = 0.00000145 \text{ psi.s}$$

Newtonian fluids. The relationship between shear stress and shear rate is most easily shown using a graph, called a *consistency curve*. Here is one curve showing how shear stress and shear rate vary for water (Fig. 7-3). Water is known as a "Newtonian" fluid when discussing its rheology. Newtonian fluids contain particles no larger than a molecule. Other Newtonian fluids are oil, glycerin, and alcohol.

In a Newtonian fluid, the shear stress is proportional to the shear rate. Viscosity equals the slope of the graph. If the shear stress doubles, the shear rate will double. In practical terms, if water is pumped through a pipe at a certain flow rate, to double the flow rate will require the pressure to double also. As soon as pressure is applied the fluid starts to move.

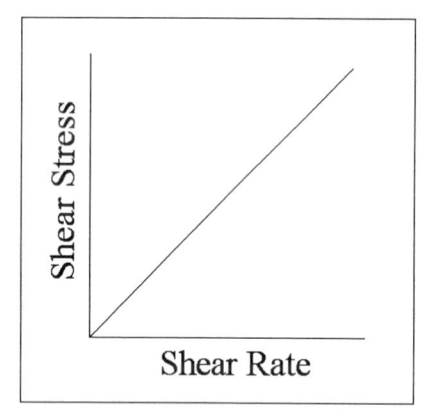

Fig. 7-3 Consistency Curve for a Newtonian Fluid

Other oilfield fluids have much more complicated rheologies. Mixtures of different fluids with any number of solids suspended within the fluids, leads to viscosities that change with the shear stress, that may change with time or that may be different depending on whether shear stress is increasing or decreasing! A fluid may turn into a gel when shear stress is removed and this requires an initial pressure to be applied before the fluid starts to move. There are many different mathematical models proposed to allow a calculation for a shear stress at a particular shear rate and some of these are very complicated. These can be collectively called "non-Newtonian" fluids.

Bingham plastic fluids. The Bingham plastic theory assumes that a fluid has a rheology where the relationship between shear stress and shear rate is linear but the line does not cross the origin of the graph. It looks something like the graph shown below in Figure 7-4.

This type of rheology exhibits a yield point. Before the fluid will flow, a certain threshold pressure must be applied and any pressure lower than this will not initiate flow. When the fluid is static, a structure builds up which connects particles in the fluid due to electrostatic attractions between them and forms a gel. For a drilling fluid, this is a desirable property because when circulation stops and a gel forms, drilled solids are suspended within the gel and do not sink down through the mud. When sufficient shear stress is applied the fluid starts to move and the gel structure breaks down again. The slope of the graph (which is a straight line in the mathematical model) is given the term "plastic viscosity". Therefore to define the behavior of a Bingham plastic type fluid, the yield point and plastic viscosity are sufficient to predict the shear stress for a given shear rate.

There is probably no fluid in existence that is a true Bingham plastic fluid that exhibits the exact rheology shown above. At the higher end of the range of shear rates present while drilling a well, it is a close approximation. The model was developed to describe the behavior of water-based mud containing clays, but is not adequate to describe the behavior of modern polymer mud.

To calculate the yield point and plastic viscosity, the shear stresses at two shear rates (511 sec^{-1} and 1022 sec^{-1}) are measured. A straight line is drawn between these two points and extended back to the Y axis. Where the line hits the Y axis it is taken as the yield point. Now the only thing that can be defined from two points is a straight line, and it is certain that in non-Newtonian fluids, this straight line relationship does not exist. The Bingham plastic model is completely useless for predicting behavior at low shear rates (where the relationship does not even approximate to a straight line) because only two measurements are taken, the lowest measurement is 511 sec^{-1}. In specifying a required rheology, performance in the annulus is the most critical because this relates to how well the hole is cleaned of cuttings by the mud. The range of shear rates in the annulus will vary between 10 and 500—exactly the range where viscosity cannot be predicted by the Bingham Plastic model.

[Figure: A graph with "Shear Stress" on the vertical axis and "Shear Rate" on the horizontal axis. A straight line begins at a point labeled "Yield Point" on the vertical axis and slopes upward to the right.]

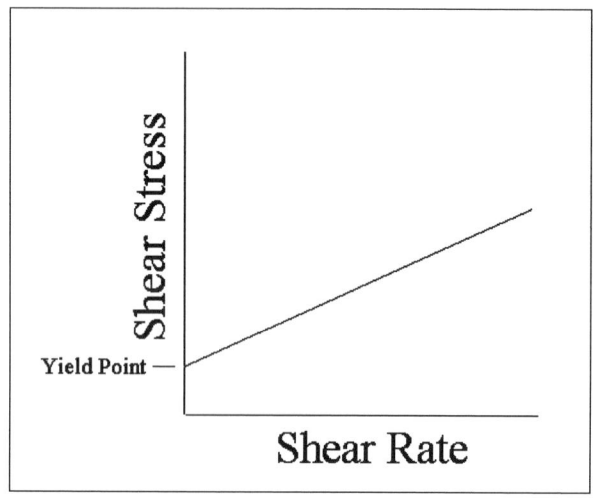

Fig. 7-4 Consistency Curve for a Bingham Plastic Fluid

Unfortunately, the Bingham plastic model is widely used to describe every type of fluid encountered in oil well drilling, even though it's almost the least useful!

The figure for yield point estimates the portion of the total viscosity that comes from attractive forces between particles suspended in the mud. The figure for plastic viscosity relates to the resistance to flow due to interparticle friction. This friction is affected by the amount of solids in the mud, the size and shape of those solids, and the viscosity of the continuous liquid phase.

Viscosity was defined as shear stress ÷ shear rate. At any point on the plot, viscosity equals the slope of a line drawn from the origin of the graph to that point. It can be seen from the graph above that as shear rate increases, the viscosity decreases. This behavior is called *shear thinning* because increasing shear causes the fluid to become thinner (less viscous). It's an important property in a drilling fluid for reasons that will be explained shortly.

Pseudoplastic fluids. Pseudoplasticity is a rheology model that demonstrates no yield point (the fluid starts to flow as soon as pressure is applied), but the viscosity decreases as the shear rate increases. After a certain shear rate, the viscosity *approximates* to a straight line. If this straight line is extended back to intersect the shear stress (vertical) axis, a point is reached that is termed *pseudo yield point*. By defining the pseudo yield point and the slope of the line, viscosities at higher shear rates can be estimated. However, as with a Bingham plastic model, this method is useless for examining low shear rate behavior. Figure 7-5 shows the consistency curve for a pseudoplastic fluid.

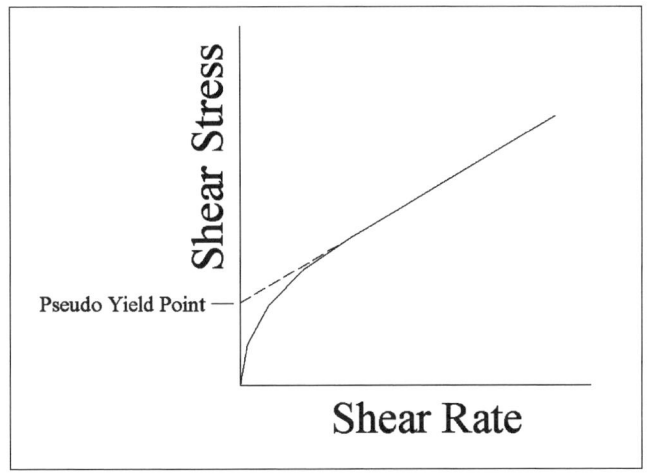

Fig. 7-5 Consistency Curve for a Pseudoplastic Fluid

Fluids that exhibit pseudoplastic behavior are typically drilling muds that contain long polymer chains suspended in the continuous phase fluid. Increasing shear rates cause decreases in viscosity due to the long polymer chains straightening out and aligning themselves with the direction of flow. Some cement slurries are also pseudoplastic over at least part of the range of shear rates experienced in cementing operations. However, cement slurries, while being pseudoplastic part of the time, also exhibit a yield point. The combination of a pseudoplastic curve with a yield point is expressed with the Herschel-Buckley model, described below.

To predict performance of a pseudoplastic fluid, a mathematical model known as the "power law" is used. This model is expressed with the following equation:

$$\tau = K\gamma^n$$

where τ = shear stress, dynes/cm^2
K = consistency index, which is a constant, equals the shear stress at 1 sec^{-1} shear rate

γ = shear rate, sec^{-1}
n = power-law index

The index "n" indicates how much the fluid diverges from Newtonian behavior; if n = 1 then the fluid is Newtonian (τ is proportional to γ). As n decreases below 1, the fluid behavior becomes increasingly non-Newtonian. If n is greater than 1, then the fluid behavior is termed "dilatent" and viscosity will increase at increasing shear rates, as shown in the curve below.

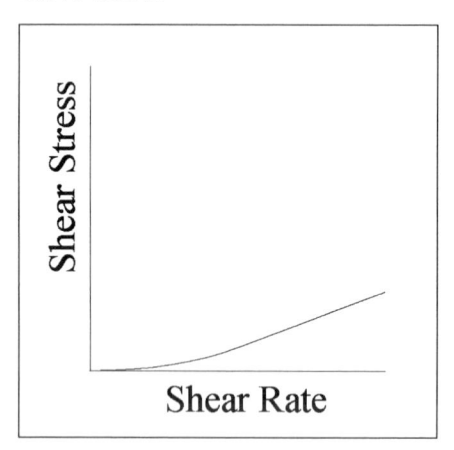

Fig. 7-6 Consistency Curve for a Dilatent Fluid

If n is lowered, hole cleaning is improved because it increases the effective annular viscosity. If K is raised, hole cleaning is improved because this, too, increases the effective annular viscosity. If, after lowering n, the hole cleaning performance is still inadequate, increasing the solids content should raise K.

Dilatent fluids. A dilatent fluid is the only one of the non-Newtonian models discussed here that does not shear thin—rather, increasing shear causes the viscosity to increase. Drilling mud should never be dilatent because this will give high parasitic pressure losses in the drill string and inadequate hole cleaning in the annulus.

Herschel-Buckley fluids. This model demonstrates a yield point, together with pseudoplastic behavior once the fluid starts to flow. Many muds and cement slurries are usefully described using this model (Fig. 7-7). The equation for this

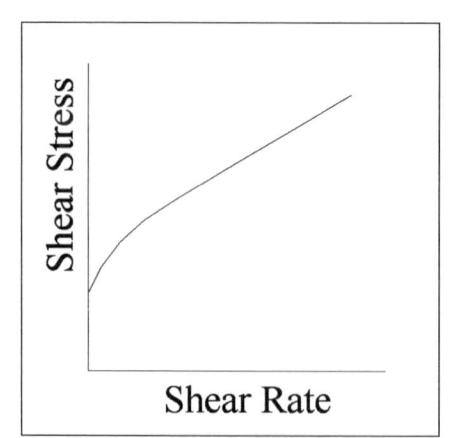

Fig. 7-7 Consistency Curve for Herschel-Buckley Fluids

model is similar to the pseudoplastic model, but a yield point is added as follows:

$$\tau = \tau_y + K\gamma^n$$

where τ_y = yield stress in dynes/cm² or lbf/100 ft²

Time dependent rheology. It is possible for physical or chemical reactions between particles in a fluid to cause a change of viscosity with time. Structures (relationships between particles) may form or break down. If the fluid is sheared at a constant rate, having first been sheared at a lower rate, the viscosity will decrease over a period of time in a thixotropic fluid until equilibrium is eventually reached. If the shear rate is decreased, then viscosity will increase over time until again equilibrium is reached. If the shear rate becomes zero (the pump is stopped) then structures will build up in the fluid and the fluid will form a gel—effectively a fluid with the characteristics of a solid.

Polysaccharides (such as guar gum and starch when mixed in water) exhibit thixotropy.

A fluid that exhibits the opposite effect, where viscosity increases with time after increasing the shear rate (and decreases with time after decreasing the shear rate), is called *rheopectic*. Muds and cement slurries that contain a lot of suspended solids (>20 pounds of bentonite per barrel of fluid, or high quantities of drilled solids) sometimes show rheopectic behavior.

With cement slurry[13], there is of course another source of time dependency because the slurry is continually undergoing chemical reactions that change the nature of the relationship between particles within the fluid as the cement sets.

Flow regimes. All of the models discussed above only deal with fluid in laminar flow. When a fluid moves in laminar flow, the fluid flows smoothly with all parts of the fluid moving in the same direction. If a fluid is pumped quickly enough, this smooth flow starts to break down. Eddies start to appear in the fluid and the flow becomes chaotic—not all of the fluid moves in the same direction. Mixing takes place due to the random flow movements within the fluid. Eventually, if the flow rate becomes fast enough, there is no laminar flow at all and the flow is termed *turbulent*. The situation where the flow is no longer completely laminar up to the point when the flow becomes turbulent is called *transitional flow*.

The shear rate at which a fluid enters turbulent flow is dependent on the dimensions of the pipe or annulus, the fluid density, and the fluid viscosity.

Rheology; design considerations. A drilling fluid in the circulating system has to pass from the mud pumps, through surface lines, down the drillpipe, through the drill bit, back up the annulus, and through the solids-control equipment before arriving back in the surface tanks. Only in two of these sections does the fluid actually perform useful work in return for the energy expended to cause flow—at the bit (to clean the bit and bottom of the hole) and in the annulus (to lift cuttings to the surface). All other pressure losses, through the pumps, surface pipes, and drillpipe, are called *parasitic*.

Flow in the lines and drill pipe will certainly be turbulent, due to the flow rates involved. As almost all drilling fluids are shear thinning, the pressure required to drive mud through the surface lines and down the drill pipe is less than if the fluid viscosity stayed constant (as it would in a Newtonian fluid). Still, the pressures involved are high and the parasitic pressure losses between the pump and the drill bit could exceed 1000 psi on a 10,000' well. Ideally, the fluid rheology will minimize these parasitic pressure losses.

As the mud passes through the drill bit, it accelerates very rapidly. The inside diameter of a drill collar in a 12-1/4" hole might be around 2-3/4"; a cross sectional area of 5.94 in². In a typical tricone bit in this hole size there will be three nozzles with an ID of around 1/2" each. Together, these three nozzles give a total cross sectional area (known as "total flow area", or TFA) of 0.45 in². If the mud is pumped at 600 gallons per minute, the flow will accelerate from 32' per second through the BHA to 427' per second through the nozzles.

One measure of the work expended at the drill bit by the mud is hydraulic horsepower (HHP). To calculate the HHP it is first necessary to calculate the pressure required to accelerate the fluid through the nozzles as

$$\text{Pbit} = \frac{\rho Q^2}{594 A^2}$$

where   Pbit = Pressure drop across the bit, psi

r = Fluid density, psi/ft
Q = Flow rate, gallons per minute
A = Total Flow Area (TFA)

For this example, if r = 0.6 psi/ft, then Pbit = 1,800 psi at 600 gpm with a nozzle TFA of 0.45 in^2.

The HHP can then be calculated by

$$HHP = \frac{PQ}{1714}$$

where   P = Pressure drop across the bit, psi
Q = Flow rate, gallons per minute

which would give an HHP of 627 for this example.

The following factors will increase the HHP if everything else remains the same

- Increase in fluid density
- Increase in flow rate
- Decrease in TFA

At the shear rates involved, the viscosity at the bit should be fairly low. A thick fluid would not easily penetrate the cracks around the rock cutting and get underneath to dislodge it. However, none of the rheology models described here are applicable at such extremely high shear rates (between 5,000 and 100,000 reciprocal seconds). What can be said is that with a shear thinning fluid, at such high shear rates with typical drilling mud, the mud will normally be thin enough to penetrate the new fractures around the cutting.

Two other functions of the mud at this point are to cool the bit and clean the bit. These are directly related to the fluid velocity, as shown above, which will normally be quite high.

To lift cuttings up the annulus, the flow rate upwards must exceed the rate at which cuttings fall down through the fluid column. The more viscous a fluid, the slower a cutting will fall down through it. For consistent cuttings transport at different flow rates, the fluid should increase its viscosity in inverse proportion to the fluid velocity—if the flow rate halves, the viscosity should double.

In the solids-control equipment, where moderate to high shear rates exist, the fluid should be thin to allow drilled solids to be easily separated out.

Drilling Technology in Nontechnical Language

Optimum mud rheology model. Now it is possible to define the optimum rheology model for a drilling fluid.

1. At extremely high shear rates (through the bit nozzles) the fluid should be very thin
2. At high shear rates (through the surface equipment and drill string) the fluid should be thin to minimize parasitic pressure losses
3. At slightly lower shear rates (in the solids-control equipment) the fluid should be thin to maximize separation of drilled solids from the mud
4. At medium to low flow rates (in the annulus) the fluid should become more viscous to give efficient lifting of drilled cuttings
5. At very low to zero flow rates, the fluid should become highly viscous to minimize settling of solids through the mud
6. The fluid should not develop excessive gel strength or else once pumping starts again after a period of rest, high pressures will be imposed on the open hole that may fracture weak formations

This would be a pseudoplastic fluid with little or no thixotropy. The power-law index "n" should be low to give good shear thinning behavior.

Rather than using rheology models to try to predict viscosities at flow rates other than those tested, it is better to run a larger number of tests at different shear rates and construct an accurate consistency curve from this data. That will allow a full understanding of the fluid behavior and allow treatments of the mud to be tailored to give specific rheology at shear rates of interest.

If a rheology model is to be used to produce a consistency curve, the Herschel-Buckley model, of those described above, is likely to give the closest match to actual mud properties and is therefore the most useful model to use in most circumstances. It can also accurately model a Newtonian fluid ($n = 1$, $\tau_y = 0$, K = viscosity).

Mud chemical properties

The chemical characteristics of the mud are mostly determined by wellbore stability considerations of the formations drilled through in a particular hole section. In addition, the mud should not damage the reservoir (reduce permeability), or at least damage done to reservoir permeability should be capable of being repaired (i.e., by using acid to remove plugging solids) or bypassed (explosive perforations penetrating through the damaged zone).

A brief description of different problems and required chemical characteristics follows.

Reactive shales. Many hole problems are caused by incompatibility between water and shales (the reverse of diagenesis, discussed in chapter 1). This may be solved by using oil/water emulsion mud (with oil as the continuous phase) or 100% oil mud. This isolates water from the shales and prevents hydration. Oil muds are becoming increasingly difficult to use in some areas due to environmental concerns and resulting government regulation changes. These muds are also expensive.

Water-based mud may use various chemical inhibitors to control reactive shales, such as potassium chloride (KCl). KCl works by changing places with sodium atoms in the clay structure and, as the potassium ion is smaller than the sodium ion, this causes the clay structure to shrink rather than expand.

Other useful chemicals include polymers. Clay crystals have electrostatic charges on their faces and edges, if a polymer molecule also has opposing charges along its length, the polymer sticks to the clay crystal and prevents water from reaching it.

A recent development is the use of soluble silicates in clay stabilization. These are soluble at high pH (alkaline), but precipitate out of solution as solids if the pH drops. As tiny amounts enter the pore spaces between crystals, the pH drops, silicate precipitates and forms a barrier to further water penetration. Use of silicates seems to cause the clay to harden over time. As silicates are cheap, readily available, and environmentally friendly, their use will no doubt increase into the future.

Salts. A non-salt-saturated water-based mud will leach out salt formations, causing extreme hole enlargement and possible cementing problems. This can be addressed by using either salt-saturated water mud or oil-based mud.

In complex salt sequences containing the most soluble potassium and magnesium salts, a mixed salt system is required to address the particular mix of salts present in the formation.

Reservoir damage. Mud filtrate can be extremely damaging to formation fluids. There are two areas of particular concern—pollution of water sources and reduced productivity of the pay zone. Two approaches may be made in these cases—either prevent the filtrate invasion by using additives to plug off pore throats where the formation is exposed and/or use a mud that has a non-damaging filtrate. In the pay zone, productivity damage from filtrate may occur in the following ways:

1. The filtrate may contain fines (small solid particles) that bridge off the zone of invasion. If this zone is deep, perforations may not be able to penetrate completely and the well will be less productive. If the fines were acid soluble, then acid treatments may remedy the situation partially or completely. However, weighting agents and drilled solids are usually not acid soluble, so fluid loss control becomes very important in the production hole section. Acid soluble materials (such as calcium carbonate) may be used.

2. Chemical reaction between filtrate and formation fluids may produce solid precipitates or blocking emulsions. As noted above, if these are acid insoluble then the resulting damage may be permanent.

3. The filtrate may react with the clays within the formation. Oil-based mud should give only low amounts of oil filtrate—no water should be present.

Corrosion of downhole steel components. Tools and tubulars used in drilling, casing and completing the well can be subject to corrosion by the mud. For most casing strings, mud is left in the annulus after the cement job, which will remain for the life of the well. Mud properties may change over time due to bacteriological action. This can produce H_2S (especially when the mud contains organic additives) or low pH levels. Oil muds produce "oil wetting" of metal surfaces and will protect against CO_2, H_2S and H_2O corrosion.

Hydrogen sulfide (H_2S) related problems. H_2S may enter the mud from a permeable formation, either as a kick or from within drilled cuttings. Apart from the extreme toxicity of this gas, it causes hydrogen embrittlement of

Chapter 7 • Drilling Fluids

most steel, which degrades tensile strength. If H_2S is anticipated, an excess level of lime in the mud will help. The (alkaline) lime will help neutralize the (acidic) H_2S. The reaction forms active sulfides (CaS, and $Ca(HS)_2$), which will liberate H_2S if exposed to a mild acid. Once H_2S has been identified in the mud system then zinc oxide (ZnO) is an effective scavenger of H_2S and active sulfide salts. This reacts to form stable zinc sulfide. It is not recommended using ZnO before H_2S is identified, as it will mask a slow entry of H_2S into the system. Recommended concentration of ZnO is around two pounds for each barrel of mud.

Chapter Summary

This chapter covered one of the most important systems on the rig for safe and efficient drilling—the drilling fluids. First, the functions of a drilling fluid were listed, then seven distinct classifications of mud were described. Drilling fluid design was covered in some detail for the physical, rheological, and chemical requirements.

Glossary

[1] **Rheology.** The study of fluids in motion. Rheological properties such as viscosity impact on several key mud functions.

[2] **Continuous phase (of a drilling fluid).** The main component of the system. The carrying phase into which everything else is mixed. (Definition from *Dowell Drilling Fluids Technical Manual*).

[3] **Polymer.** A repeating chain of units (called *monomers*) that are chemically joined together to form a long chain. Some polymers can be millions of units long.

[4] **Viscosity.** The degree of resistance to flow of a fluid. A highly viscous fluid will need more pressure to pump it through a pipe than a fluid with lower viscosity.

[5] **Gellation.** When the fluid becomes stationary, it forms a gel. This allows solids in the mud (including drilled cuttings) to be suspended within the gel so that they don't fall down through the mud.

6. **Fluid loss.** A mud containing solids that is in contact with a permeable formation will lose water and dissolve chemicals into the formation. As this happens, solids in the mud will form a plaster or cake on the face of the formation. This loss of liquid (called *filtrate*) into the formation is called fluid loss. Some chemicals can reduce the amount of fluid lost to the formation in this way.

7. **Workover.** The process of repairing damage to a well, often (but not always) by removing the completion and running a new one.

8. **Differential sticking.** Where

   - permeable formation is exposed in the wellbore
   - the mud hydrostatic pressure is greater than the pore fluid pressure
   - mud solids have built a plaster on the formation face
   - the drillstring is allowed to remain stationary for a while

   It is possible for the drillstring to become stuck. The greater pressure in the wellbore pushes the pipe into the wall and friction then makes it hard to move the pipe up, down, or in rotation. If the overbalance is high enough, the pipe cannot be moved. Reducing the drilling fluid density reduces the sticking effect.

9. **Circulating pressure losses.** This refers to the pressure necessary to force fluid to flow along a pipe or annulus. The pressure at the bottom of the well while circulating equals mud hydrostatic pressure plus the pressure required to force the mud to flow up the annulus. This extra pressure imposed while circulating along the open hole can be enough to fracture weak formations in some cases.

10. **Filtrate.** The liquid part of the mud. When mud is forced against a permeable zone, the solids in the mud form a plaster or "wall cake" against the formation face. Some of the liquid fraction will filter through this cake and into the formation. This liquid fraction (water plus dissolved salts) is called *filtrate*.

11. **Surge and swab pressures.** When the drillstring is lowered in the hole, fluid is displaced upwards. This imposes temporary extra pressure on the

Chapter 7 • Drilling Fluids

hole and is called "surge" pressure. When the drillstring is lifted upwards, fluid has to flow downwards as a pressure drop is created by the withdrawal of the steel volume. This causes a temporary pressure reduction on the hole and is called *swab* pressure.

[12] **Hole cleaning ability.** The ability of a drilling fluid to lift cuttings out of the hole at a certain flow rate. This ability is related to the fluid density and rheology.

[13] **Slurry.** A suspension of solids in water. A cement slurry normally consists of a mix of water, various chemicals, and cement powder. Cements are discussed in detail in chapter 9.

Chapter 8

Directional and Horizontal Drilling

Chapter Overview

Directional drilling—the process of accurately guiding a well through a predefined target or targets—is a source of great interest for many people outside the drilling industry. This chapter will first give some of the reasons for drilling directional wells. The tools and techniques for deviating the wellbore will be described. Accurate navigation is a pre-requisite to drilling directional wells, therefore the many tools available to the driller, along with their advantages and limitations, will be covered. Finally, this chapter will discuss some specifics concerning drilling horizontal and multilateral wells.

Why Drill Directional Wells?

There are several reasons why the extra cost of drilling directional wells might be justified. The decision almost always comes down to simple economics—in terms of developing a field for commercial exploitation, directional drilling gives the highest return on investment.

Single surface location

In some cases (most offshore production wells), many wells are drilled from a single location on the seabed to many downhole target locations. These targets will be selected to give the best overall production from the field. As the wellheads are all gathered together at the surface, these wells can easily be hooked up to production facilities.

Drilling Technology in Nontechnical Language

Traditionally a platform would be installed offshore, and permanently fixed to the seabed. Fixed platforms are expensive to build. Once production is finished and the field is abandoned, fixed platforms are also expensive to remove. Some interesting alternatives to fixed platforms are now used that allow the surface facilities to be moved to a new field after abandonment.

Tension leg platform (TLP). A floating platform is built that is anchored to the seabed using steel pipes. Once the pipes are in place, jacks on the platform pull on the legs so that the floating platform is pulled down in the water. This exerts a tension on the pipes. In this way the platform does not move up and down with tides or waves. When the field is abandoned, the TLP can be released and moved somewhere else. A TLP can be used in very deep water—much deeper than traditional fixed platforms.

Floating production and storage offshore (FPSO[1]). A ship with production facilities on board is permanently moored in the field. The wells terminate at a template on the seabed and hoses then route the production up to the ship, which treats the produced oil and stores it ready for collection. Shuttle tankers call at the ship and offload the stored oil.

Production buoy. A floating unmanned buoy takes production from a subsea manifold and stores the oil for shuttle tankers to offload.

Jackup production platform. Jackup rig technology is used to build a platform that is moved on location and jacked up on legs so it is clear of the sea surface. Once production is finished, the platform is jacked down into the water and moved off. Only suitable for shallower waters (less than about 300' water depth). Some of these platforms are converted jackup drilling rigs with the drilling facilities removed.

Inaccessible location

Sometimes the desired target is below a surface feature that prevents drilling a vertical well. Natural features might make it difficult to place a rig on a location vertically above the target (such as a swamp or environmentally sensitive site) or artificial features such as buildings or an airport. In this case an alternative location can be chosen that allows the target to be reached with a directional well.

Chapter 8 • Directional and Horizontal Drilling

Salt dome drilling

As described in chapter 1, salt domes can provide good traps for hydrocarbons. Drilling through the salt dome itself can bring some very difficult drilling problems. The hydrocarbons could be exploited instead by directional wells that drill around the salt instead of through it.

Multiple exploration wells from a single wellbore

If several interesting structures lie close together, an exploration well could be drilled so that after drilling into one structure, the well could be cemented back a distance and sidetracked to another structure. This could be cheaper than drilling a separate well to each target.

If several different structures lie vertically stacked up, a well could be directed to drill through several or possibly all of them. Sometimes, a large fault could create several traps below the fault and by drilling a well alongside the fault, all the structures along the fault that might contain hydrocarbons can be penetrated.

Onshore drilling to an offshore reservoir

Where a reservoir lies close to the land, it may be possible to access the reservoir by drilling from a land rig. Wells drilled on land are invariably cheaper than drilling and producing from an offshore location.

Horizontal well in reservoir

Some reservoirs have good vertical permeability but low horizontal permeability. An example would be a fractured limestone reservoir where the fractures are vertical. If a well were to penetrate many fractures, it would be able to produce far more oil than a vertical or low angle directional well that only penetrated one or a few fractures.

Remedial work (sidetracks)

A sidetrack is drilled when an existing wellbore is cemented back and a new wellbore drilled away from the original wellbore. Sidetracks may be drilled for many different reasons. For instance, if the drillstring became stuck downhole and could not be freed, it could be cut downhole using explosives and cement placed on top of it. Then the well could be drilled directionally around the original wellbore and back towards the target.

Relief wells

This is the worst time to drill a directional well! If a well starts to blow out and cannot be contained from the well surface location, then it may be necessary to drill a directional well to intercept the blowing wellbore. Once communication is established with the blowing well, heavy drilling fluid is pumped down the relief well and into the blowing well to kill it.

Tools and Techniques for Kicking Off[2] the Well

There are three methods of accurately kicking off the well—jetting, whipstock, and downhole motors. The motor techniques are most commonly used because they are fast and accurate. However, the whipstock is still used. Jetting is rarely used, but it's still a valid and inexpensive technique.

Jetting

This is an old, "low tech" technique that can still be worth using in suitable circumstances.

A drill bit using rotating cones to cut the rock usually has three cones. Between the cones are nozzles that direct the flow of high-pressure mud past the cones and to the bottom of the hole. The hydraulic energy expended at the bit works to clean the bit and also the bottom of the hole. Ideally the bit's hydraulic configuration will remove cuttings from the bottom and from the cones so that each time a tooth contacts the bottom of the hole, it will touch and cut virgin rock and not re-drill cuttings from a previous tooth contact.

The nozzles are tungsten carbide inserts with a standard sized outside diameter (to fit in a standard nozzle pocket) but the inside diameter can be varied. If one nozzle is sized to be very large (13/16") and the other two are quite small (5/16") then most of the flow will pass through the large nozzle.

At high flow rates in softer formations, this large flow can be enough to wash away the formation on that part of the hole where this flow is directed. If the drill bit is then aligned so that this large nozzle points in the direction that the well should be deviated, then a pocket is washed away on that side and the well will deviate in that direction. This technique is called *jetting* or *badgering*.

One bit company manufactured a special bit for badgering. It was a three cone bit with the third cone missing and instead of a third cone, a large mud port was placed in the third cone position.

Badgering is used to kick off a well in softer formations close to the surface. As it doesn't involve expensive downhole motors it can be a very cost effective method of initially deviating a well.

Whipstock

Another old method of deviating the well (which is still used) is the whipstock. This is a wedge with a concave face that is placed in the well. The face of the whipstock is pointed in the direction the well must go. Once the whipstock is set on the bottom, the drill bit starts to rotate and drill away rock on the side of the wellbore. As the bit starts to drill and moves lower down the face of the whipstock, the wedge-shaped profile of the whipstock forces the bit further into the formation.

After drilling 15-20 ft. of rock, the drill bit is pulled out. The whipstock comes out hanging on the bottom hole assembly. A new flexible drilling assembly is run in and more of the hole is drilled. This also smoothes out the ledge left behind under the whipstock (Fig. 8-1).

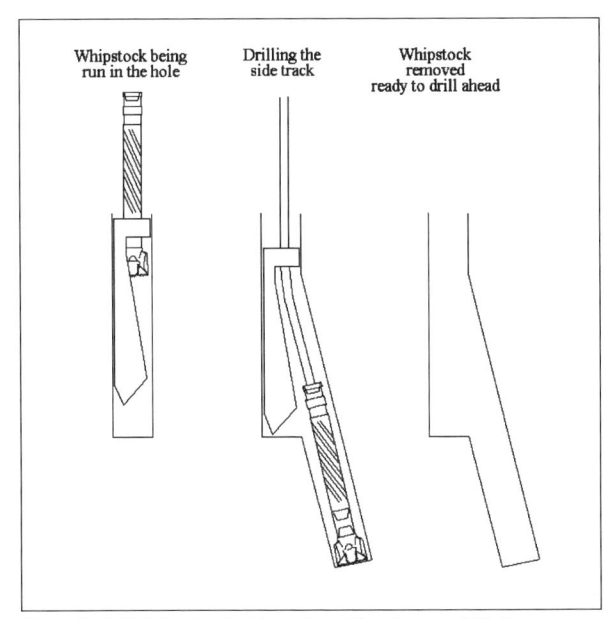

Fig. 8-1 Whipstock Used to Deviate a Well

Drilling Technology in Nontechnical Language

Downhole motor and bent sub

A downhole motor looks somewhat like a drill collar on the outside. Inside the motor is a mechanism that converts hydraulic horsepower (from drilling mud being pumped through the motor) to rotary power. The drill bit is made up on the bottom of the motor. With the drillstring stationary and the mud pumps on, the bit turns and will drill.

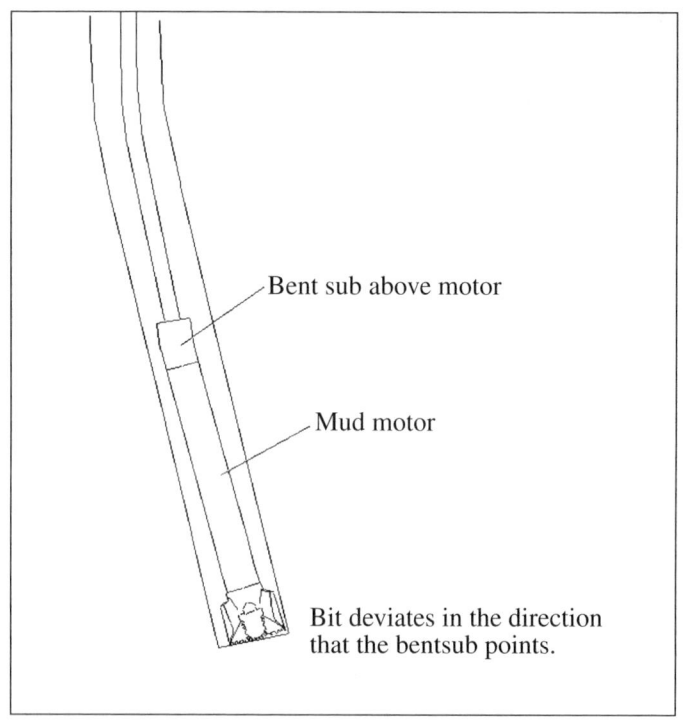

Fig. 8-2 Bent Sub Above Motor

Above the motor is a short sub, which has a slight bend in the middle. Of course, this is called a *bent sub*. The well will deviate in the direction that the inside of the bent sub points to (Fig. 8-2).

A downhole motor can also be run without a bent sub. This might be done in circumstances where it is preferred not to rotate the drillstring to drill ahead—such as if wear on the casing due to the rotating drillstring might be a problem.

Downhole steerable motor

A steerable motor has a bend close to the bottom end—close to the drill bit. As the bend is in the outside of the motor—the housing—it's termed a *bent housing*. The advantage of a steerable motor over a motor with bent sub is that the steerable motor can drill straight, if the drillstring is rotated. As the bend is close to the drill bit, this will not cause undue forces on the motor (as it would if a motor and bent sub were rotated in the hole). So now one drilling assembly can be run that will drill straight if rotated, or will deviate if the drillstring is kept stationary.

A steerable motor is more expensive than a downhole motor and bent sub. It will be used in preference to a bent sub system if the steerable capability is required to drill a precise directional path, or if the ability to both deviate and drill straight without tripping is important.

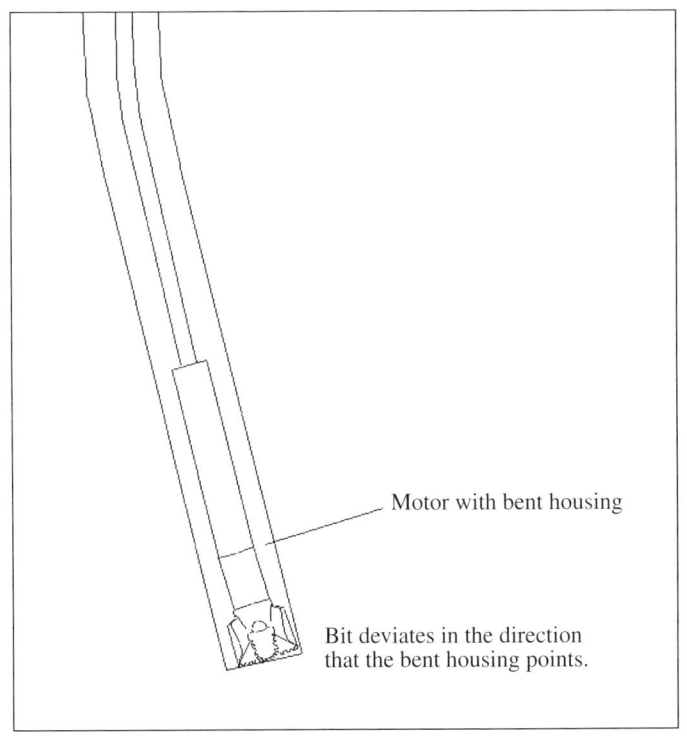

Fig. 8-3 Downhole Motor with Bent Housing

Controlling the Wellpath of a Deviated Well

Once a well has some inclination away from vertical, there are other tools and techniques that can be used to change direction or inclination. While jetting is only used for an initial kickoff, the downhole motor techniques can be used to build or drop inclination or to change direction.

Rotary drilling assemblies can be designed to increase or decrease inclination, to turn the well left or right, or to maintain the existing direction of the well. Rotary assemblies—which work while the drillstring is rotated—are less likely to get stuck than motor assemblies that have to slide while drilling in order to deviate in a particular direction . Rotary assemblies are also cheaper (because expensive motors are not required), less trouble (no motor to quit), and usually allow faster progress (restricted drilling parameters and hydraulic power at the drill bit due to the motor).

Stabilized assemblies

Stabilizers were described in chapter 3. A stabilizer is like a short drill collar that has blades extending out to touch the wall of the hole. By positioning the stabilizers along the BHA and selecting either full gauge stabilizers (that have blades extending out to the same OD as the drill bit) or undergauge stabilizers (that have blades extending out to an OD less than hole diameter), the BHA can be designed to drop angle (become more vertical), hold angle (follow the same direction), or build angle (increase the angle from vertical).

Hold , tangent, or locked assembly. This BHA typically has a full gauge stabilizer just above the bit, then a drill collar followed by another full gauge stabilizer, and a second drill collar with another full gauge stabilizer above that. These three full gauge stabilizers positioned around 35' apart, together with the stiffness[3] of the drill collars, will keep the hole in the same direction.

A hold assembly allows maximum drilling parameters (weight on bit and rotary speed) to be used, so if the well is pointing in the desired direction (even if this is vertical) then a hold assembly should be used so as to allow maximum drilling progress.

Drop assembly. The drill bit has above it two or three drill collars and then a stabilizer. The stabilizer holds the BHA at that position (60' to 90' above

Chapter 8 • Directional and Horizontal Drilling

the bit) in the center of the hole. The drill collars below it, being heavy and long, will tend to lie on the low side of the hole. This exerts a side force on the drill bit so that as it drills, it will try to cut the hole towards vertical. The main problems with a drop assembly are

1. Because it's so flexible, control of azimuth (direction) is difficult
2. As the well gets closer to vertical the rate of dropping angle decreases
3. Low weights on the drill bit are necessary so drilling progress will be slow

Placing high weights on the bit will make the drill collars buckle unpredictably, which might make the bit build, turn, or drop.

Build assembly. The drill bit has a stabilizer immediately above it, then two drill collars before the next stabilizer. This 60' of drill collar between stabilizers will tend to sag in the middle, which will lever the drill bit so it presses against the upper side of the hole. The sagging (or buckling) will increase as higher weights are placed on the bit, so drilling progress can be fast because of these high weights.

Rebel tool

There is an interesting tool that was used in the past to give a gradual turn to the left or the right. It is positioned immediately above the bit. It works by using gravity to push against a lever, or "paddle", at the upper end of the tool. When this paddle reaches the low side of the hole (as the BHA rotates) and the weight of the BHA pushes down against it, this leverage force is transmitted along a connecting rod to another paddle at the bottom end of the tool. In Figure 8-4 below, this bottom end paddle then pushes against the right hand side of the hole, pushing against the wall of the hole and exerting a side force on the bit to make it drill to the left. So once on each revolution, the bottom paddle pushes against the right hand side of the hole.

The rebel tool will give a long gradual deviation to the left when long paddles are used.

There is also a set of short paddles. In this case, the bottom paddle pushes against the left hand side of the hole, forcing a gradual turn to the right, around 1°-2° of turn per 100' drilled.

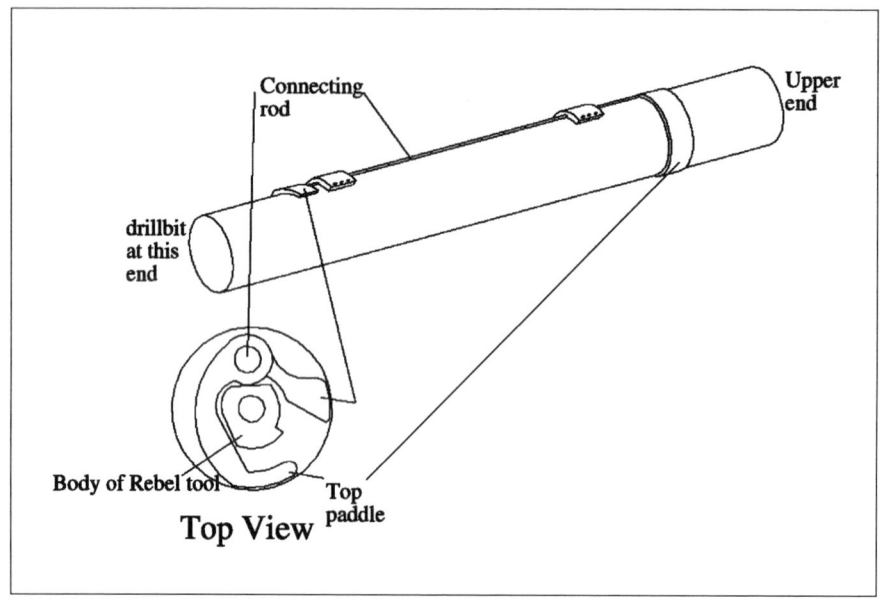

[Figure: Rebel Tool conceptual drawing with labels: Connecting rod, Upper end, drillbit at this end, Body of Rebel tool, Top paddle, Top View]

Fig. 8-4 Rebel Tool Conceptual Drawing

More recently, modern alternatives to the rebel tool have come into use. Rebel tools are hardly ever seen in modern drilling.

Andergauge mechanical steering tool

This tool, manufactured by Andergauge, cleverly uses a long weight that will tend to swing to the low side of a mandrel passing through it. The mandrel rotates to turn the bit that is below the tool. The weight is connected to a paddle that exerts a force on the side of the hole. The position of the paddle relative to the weight can be adjusted while the tool is down the hole, therefore the direction the tool steers the bit can be changed as required.

Baker Hughes "Autotrak"

This ingenious tool uses downhole computer and hydraulic systems to dynamically control pistons in stabilizer blades, so that as each blade passes the direction opposite to that desired for the bit to drill, the pistons move out and push against the wellbore. This tool can be set to measure downhole rock properties and use those measurements to steer automatically inside the reservoir. A very high tech solution!

Horizontal Wells

Horizontal wells have been drilled with incredibly long horizontal sections—measured in kilometers. The technology has evolved to allow horizontal drilling to become almost routine.

The following problems are associated with drilling horizontal wells:

- **Hole cleaning.** As the drillstring lies on the low side of the hole, beds of cuttings build up around the bottom edge of the drillstring. These can be very hard to shift.

- **Frictional forces.** The power needed to turn the drillstring or to pull it out of the hole are higher on a horizontal well than on a normally deviated or vertical well.

- **Accurate navigation in the reservoir.** Navigation within the reservoir is relative to reservoir characteristics and not computed according to inclination and azimuth only.

Many directional wells are drilled with two built up sections, as was described in chapter 4. Figure 8-5 shows the different sections of a typical horizontal well.

Almost all horizontal wells are drilled as production wells (they cost significantly more than vertical or moderately deviated wells) and if all the wells come back to the same surface location, then most of the targets in the reservoir will be displaced horizontally some distance away from the surface location. This makes it necessary to have a first build, then a tangent (straight) section before finally building inclination to horizontal.

For the kickoff and first build, typically a downhole motor and bent sub will be used, as this is more cost effective than a steerable system. However, if the ability to be able to actively steer the well around obstacles (such as other previously drilled wells) was important, then a steerable system would be used.

The tangent section can be most easily and cheaply drilled with a rotary drilling assembly, with three full gauge stabilizers placed close together above the drill bit.

When the well reaches the second build, it is necessary to be able to drill very accurately to place the horizontal wellbore at the right depth. This normally requires the steerable motor. If the well ends just a few feet too

Drilling Technology in Nontechnical Language

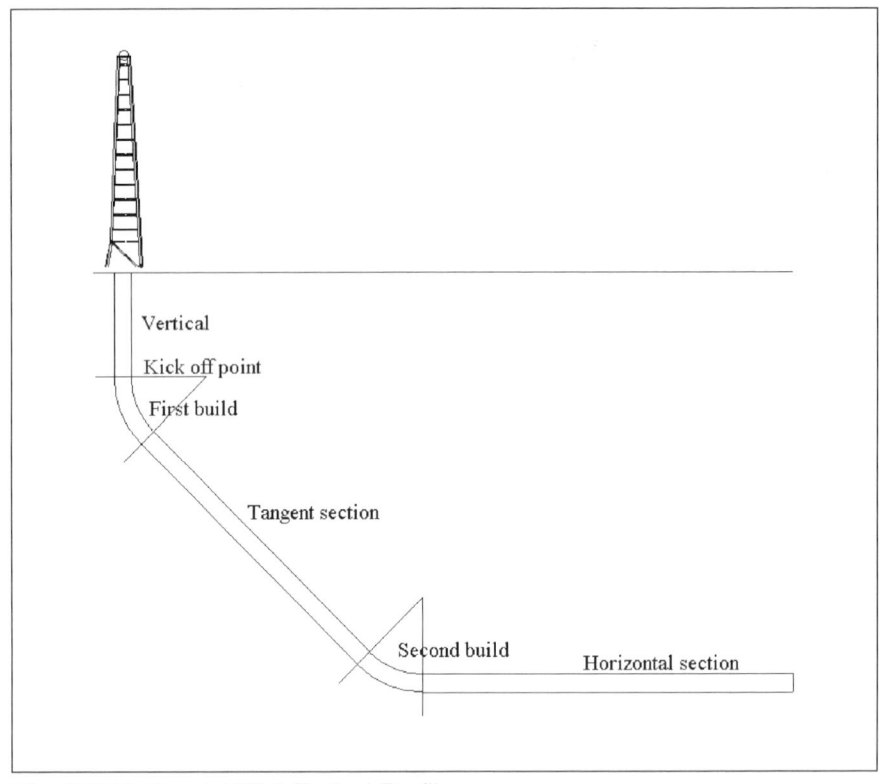

Fig. 8-5 Horizontal Well Typical Profile

low or high, it can take a few hundred feet of drilling to get back into the correct position in the reservoir so it's really important to get it right the first time. Also if the well is too low rather than too high, the BHA has to drill at an inclination of more than 90°—drilling uphill, if you like. While this is possible, it's undesirable for various reasons.

Generally, horizontal wells do not have liners cemented inside the reservoir. At the end of the second build (with the well in the reservoir, but before drilling the horizontal section) the production casing would be run. This is done to keep the reservoir isolated from the formations higher up. With the casing cemented into the reservoir, drilling horizontally can continue without having to worry about the formations above the reservoir.

Chapter 8 • Directional and Horizontal Drilling

Multilateral Wells

Multilateral wells are a relatively recent development. Several "branch" wellbores are drilled from a primary "trunk" wellbore (Fig. 8-6). This can be done for several reasons

- To place wellbores in several different reservoirs
- To get increased production in one reservoir

Multilateral sections in layered reservoirs

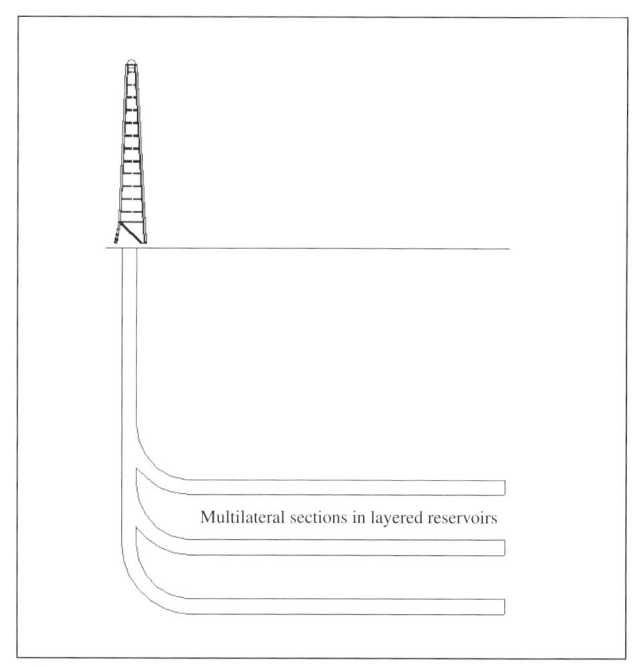

Fig. 8-6 Multilateral Well Diagram

Why is it an advantage to place several branches of the wellbore into one or more reservoirs? There are many applications now emerging as the technology matures.

- To place wellbores in several thin stacked reservoirs that would be uneconomic to exploit with separate horizontal wells
- To intercept as many fractures as possible in fractured chalk reservoirs

Drilling Technology in Nontechnical Language

- To intercept many pockets of hydrocarbons in highly faulted reservoirs
- To increase production rates and ultimate recovery volumes of oil from a single well, to either reduce the number of wells needed or to speed up field exploitation
- To increase exposure to the reservoir, decreasing drawdown pressures[4], and delaying coning
- To re-use an older wellbore that would otherwise be abandoned

One well in North Louisiana has 10 branches from one horizontal well, with the branches extending between 500' and 1,000' in length into the reservoir from the trunk! This well was drilled in 1988 by Gardes Directional Drilling, one of the early pioneers of the technology.

Surveying

In order to guide a wellbore to a desired target, the position and direction of the wellbore at any particular depth must be known. Since the early days of drilling, various tools have been developed to measure the inclination and azimuth of the wellbore.

To calculate the 3D path of the wellbore, it is necessary to take measurements along the wellbore at known depths of the inclination (angle from vertical) and azimuth (direction normally relative to true north). These measurements are called *surveys*.

To compute the wellpath between two surveys, various mathematical constructions have been proposed. The most common modern method assumes that the wellpath forms a perfect arc between two surveys. This model is called *minimum curvature*. The calculations can then be done to calculate the position of the second survey if the position of the first survey is known.

There are errors inherent in these calculations—first, survey instruments are only accurate to a certain degree, and second, though a perfect arc is assumed, this is unlikely to be the case. The calculated wellpath will be more accurate if surveys are taken close together (every 20' – 30') and if the well is not highly deviated.

One of the earliest tools used to document the wellpath was a bottle of acid that was lowered down the well. After half an hour, the acid would etch

Chapter 8 • Directional and Horizontal Drilling

a mark on a copper cylinder inside the bottle. When the bottle was pulled back to the surface, the inclination of the well at the depth where the bottle was left could be measured by measuring the angle of the etched mark.

The acid bottle could be called a type of "single shot" tool because it only takes one measurement each time it is run in the well. There are two other single shot tools that are still used today, the acid bottle having been consigned to the history books.

The first of these tools is called a *totco* tool. It is a mechanical tool that takes a reading showing the inclination of the wellbore, but not the azimuth. It is used in vertical wells to check that the well is within a few degrees of vertical. While drilling, the tool can be run on a wire to the BHA and recovered as soon as the survey is taken. A clockwork timer determines when the survey is taken. When the drillstring is tripped out, the totco survey tool can be dropped down the drillstring (with the timer set to give enough time to fall to the bottom) and will be recovered when the BHA is back at the surface. The survey is recorded by punching a hole in a paper disk.

The second single shot tool is the *magnetic single shot* (MSS). This tool measures both inclination and magnetic azimuth. The azimuth is normally converted to true north before being used in calculating the well path. For this tool to measure azimuth, it cannot be placed inside a normal steel drill collar. Normal drill collars have their own magnetic field that renders the azimuth reading magnetic compass unreliable. A special steel compound called "Monel" is used to make non magnetic drill collars and stabilizers, which have to be placed in the BHA to allow the MSS to be used. In a directional well, MSS surveys will be taken more frequently when the well is being deviated (in the build up sections) and less frequently when the well direction is not being changed intentionally (in the tangent section). The MSS has interchangeable measuring units that allow inclinations up to 90° to be recorded onto a film that is developed on the surface.

A MSS can also record the orientation of the tool face. When the well is kicked off using a jetting assembly or a downhole motor, the assembly can be orientated in the correct direction by running a MSS to see the tool face azimuth, then the drillstring can be turned at the surface to correctly align the tool face.

The totco and MSS surveys are routinely used while drilling with rotary assemblies. There are also survey tools that allow surveys to be taken in "real time" and display the data to the driller. These are called *measurement while drilling* (MWD) survey tools.

Drilling Technology in Nontechnical Language

Running surveys on wireline takes time and involves some risk. If the line were to break and the tool dropped down the well with wireline behind it, the string would have to be pulled out of the hole to recover the tool and broken line from inside the drillstring. Keeping the string stationary while taking such surveys also gives an increased risk of the pipe sticking.

In a vertical well, it's only necessary to ensure that the well doesn't suddenly head off somewhere else. There are tools that can be made as part of the drilling assembly; measuring the inclination and then transmitting this information back to the surface. One such tool is the *"Anderdrift vertical inclination indicator"*, made by Andergauge. When a survey is required, the mud pumps are stopped for 90 seconds during which time the string must not be moved. When the pumps are restarted, the tool causes temporary restrictions inside the drillstring, which can be seen as an increase in pressure on the drill floor pressure gauge. At zero degrees, 11 pulses are seen at the surface and for every half degree of inclination, one pulse less is generated. So if seven pulses are counted, the tool has measured 2° of inclination. This tool allows fast and risk free surveys whenever desired and pays for itself by significantly reducing the time needed for taking surveys.

Many tools use mud pulses to transmit information to the surface. The Anderdrift has a slow transmission rate—one pulse every 2 to 4 seconds—and the driller counts the number of pulses by watching the pressure gauges. Other tools have much higher transmission rates to allow all sorts of information to be presented to the engineers. At these transmission rates (up to 50 pulses a second), a computer is required to decode the information.

Computerized measuring devices are built inside a heavy walled tube, the same diameter and length as a standard drill collar. These tools can measure azimuth, inclination, and toolface azimuth. It is also possible to add sensors to detect and transmit data on drillstring vibrations, actual weight on the drill bit, and other information of interest.

Transmission can be made using mud pulses as described above. There are also MWD tools that are permanently connected to a surface computer by using an electrical wireline. These can only be used when the drillstring will not be rotated, but they do give the highest transmission rates.

So far the tools discussed are used while drilling. After drilling a hole section, the wellpath calculated from the MWD or MSS surveys might be confirmed by running a survey inside the drillpipe before pulling out with the last drill bit, or by running a survey inside the steel casing once it is cemented in place. These are multishot surveys (as opposed to single shot surveys); the tool records a sequence of surveys on to a strip of photographic film.

With multishot surveys, there are currently two methods of measuring the azimuth—either using a magnetic compass (magnetic multishot, MMS) or a gyro compass (gyro multishot, GMS). An MMS must be run by placing it inside a Monel drill collar, otherwise the azimuth could not be recorded. The GMS can be run inside steel drill collars or steel casing, as the gyro does not need to measure the earth's magnetic field.

The multishot instrument has a timer to take a photographic record of the azimuth and the inclination at defined intervals, usually one, two or three shots per minute. Some instruments have fixed intervals, others can be set before deployment. The engineer in charge of the survey keeps a record of the time and the depth of the survey instrument. When the film is developed and the surveys viewed, they are tied to the depth of the survey by the time.

In deep wells, the downhole temperatures can be pretty high. This will reduce the power of the batteries and can also ruin the film. For these hot surveys, a heat shield is placed around the tool to slow down the rate at which the tool internals heat up. However, if the tool is left in high heat for too long, it will eventually burn the film.

Navigation by Reference to Reservoir Characteristics

There are many tools for measuring directional information and these are vital to get the well in the right place within the reservoir. These measurements only give position relative to the wellhead, which will be fine for most wells. With horizontal wells, or other wells that require very accurate positioning of the wellbore, navigation must also be made with respect to reservoir characteristics.

MWD tools were mentioned earlier in this chapter. Logging was mentioned in chapter 3. Both of these technologies measuring formation characteristics and real-time data transmission to the surface are combined in *logging while drilling*, LWD.

Downhole sensors can measure a wide variety of rock properties in real time while drilling. By understanding the reservoir and its physical characteristics, the wellbore can be placed with extraordinary accuracy in relation to the reservoir.

One of the main problems when navigating in the reservoir is that it takes so long to change direction. The measuring instruments are at some distance behind the drill bit. A change is detected that indicates that the well has to change direction—up or down, left or right, or some combination of these. A steerable bottom hole assembly allows the wellpath to be changed in the desired direction, but it takes time (distance drilled). It's like steering a very large and heavy ship through a narrow channel.

Chapter Summary

This chapter examined some of the tools and techniques used to drill a well along a defined path. Some of the following reasons for drilling directional wells were mentioned

- To drill many wells from a single surface location
- To exploit a reservoir as effectively as possible

Next, the process of kicking off a well was described, then how to continue to guide the wellpath after the kickoff, including rotary and motor assemblies.

Horizontal and multilateral wells are more recent developments in directional drilling. These wells show the technological state that the process has now reached.

Finally, this chapter examined the techniques of surveying the wellbore and the main tools in current use.

Glossary

[1] **FPSO.** Floating production and storage offshore. Often a converted tanker, but may be purposely built. Contains production equipment to treat hydrocarbons and storage tanks to store the treated crude until a tanker arrives.

[2] **Kick off.** The act of deviating a well from vertical is termed "kicking off".

[3] **Stiffness.** Resistance to bending forces.

[4] **Drawdown pressure.** The difference in pressure between the reservoir and the wellbore when a well is in production.

Chapter 9

Casing and Cementing

Chapter Overview

An oil or gas well is a pressure vessel– a pipeline that conducts hydrocarbons from a reservoir to the surface. The integrity of this pressure vessel comes from the steel pipe that lines the wellbore. This casing pipe has to maintain integrity for the whole production life of the well, until the day it is finally abandoned. That could be 40 years or more. It is vitally important that the casing itself is properly designed, considering all the forces and environmental factors that it will be subjected to.

Outside of the casing cement is placed. The cement has to support the casing (the physical loads) without long term deterioration. It has to protect the casing from corrosion due to salt water within the formations around it. The cement must also prevent formation fluids and gases from moving up the annulus to cause problems in higher formations or possibly to affect the integrity of the surface facilities.

Cement is also used for other purposes during drilling a well. It may seal off zones that allow mud to leak into the formation. It is used to abandon a well by sealing the wellbore to prevent fluids and gases from migrating to the surface. It is often used to seal off the lower part of the well and allow a new hole to be drilled away from the old wellbore.

Casing and cementing will be described so as to impart an understanding of the importance of the casing and cement, how these things are designed, and how they are placed in the well.

Importance of Casing in a Well

Casing provides different functions during drilling, completing, and producing a well. In a deeper well there may be half a dozen different kinds of casing used to perform the necessary functions at different stages of drilling and completing the well.

Conductor pipe

The first casing is usually called the conductor. It may be driven into the ground with a pile driver or it may be cemented inside a drilled hole. The conductor is not set deep into the ground, so there is no strength to hold formation pressures in the event of a kick. The purposes of the conductor are to:

- Conduct drilling fluid returns back up to the rig during surface hole drilling so that a closed circulation system can be established. A closed circulation system is desirable so mud returns can be treated, drilled solids can be removed and re-used. (An open circulating system is where the fluid returns from the well are lost, for instance, into the sea).
- Protect unconsolidated surface formations from being eroded away by the drilling fluid.
- Sometimes support the weight of the wellhead and BOPs.

Surface casing

The surface casing is the first casing that is set deep enough for the formations at the shoe to withstand pressure from a kicking formation further down. The purposes of the surface casing are to

- Allow a BOP to be nippled up so that the well can be drilled deeper
- Protect freshwater sources close to the surface from pollution by the drilling fluids
- Isolate unconsolidated formations that might fall into the wellbore and cause problems

Intermediate casing

A shallow well may not need an intermediate casing; a deep well may need several. The intermediate casings serve as staging posts between the surface casing and the production casing. The primary purposes of the intermediate casings are to

- Increase the pressure integrity of the well so that it can be safely deepened
- Protect any directional work done (e.g., kicking off a directional well is often done under surface casing and is then protected by the first intermediate casing)
- Consolidate progress already made

Production casing

This is the long term pressure vessel. The production casing houses the completion tubing, through which hydrocarbons will flow from the reservoir. If the completion tubing were to leak, the production casing must be able to withstand the pressure.

Sometimes the production casing is cemented in place with the casing shoe above the reservoir and another hole section drilled. This may then be protected with a liner rather than a string of casing. A liner is effectively a casing that does not extend to the surface but ends somewhere inside the production casing. There are pros and cons to liners.

Advantages include:

- Economics. The cost of the liner and associated equipment is less than the cost of a full string of casing to the surface
- Utility. The inside diameter of the liner is inevitably less than the ID of the production casing. This allows tools to be run as part of the completion that would be too large to fit inside the liner but could be set higher up, inside the casing

Disadvantages include:

- Complexity. The equipment required to run a liner is much more complex than for a casing so there is more chance that something will go wrong

Drilling Technology in Nontechnical Language

- Obtaining a good cement job. Cement volumes tend to be pretty small around liners so a bit of contamination of the slurry by drilling mud will go a lot further

If a casing or liner is run through the reservoir and cemented, the casing is perforated using shaped explosive charges. These charges create a tunnel through the casing and may continue up to a couple of feet inside the reservoir. If the casing penetrates several hydrocarbon bearing zones, it would be possible to perforate the lowest zone, flow it until it's depleted then cement it off and perforate a higher zone.

Designing the Casing String

When a drilling engineer has to design a set of casings for a well, there are quite a few considerations to be made. First it is necessary to predict all the physical forces that each casing may be subjected to throughout the life of the well (from starting drilling until the well is finally abandoned).

The chemical environment also must be understood—sometimes corrosive fluids are produced from the reservoir. This will lead to special steel alloys being used, which tend to be expensive and sometimes difficult to handle when running into the hole. When corrosive pore fluids are present in formations penetrated by the well, the outside of the casing can be attacked and corroded. The cement has to form a protective barrier around the casing.

The following are main casing design considerations:

Tension. Each piece of casing will be in tension from the weight of the casing below it. The tension will therefore increase from the bottom to the top.

Downhole tubulars (casing, tubing, and drill pipe) are specified by "weight per foot" as well as other attributes. This might seem a strange way to specify a casing, but casing design tables will give, for each casing OD, a choice of "weight" and also of "grade" (which is the type of steel alloy used). A common production casing size is 9-5/8" OD and this comes in weights-per-foot of 36.0, 40.0, 43.5, 47.0, 53.5, and 58.4 lb/ft. Neglecting buoyancy forces for a moment, the weight of a 10,000 ft. string of 9-5/8" 53.5 lb/ft casing would be 535,000 lbs, which is more than 242 tons. A long string of casing can be pretty heavy!

Additional tensile forces are imposed on the casing. In a deviated well where the casing has to bend, the tensile stress in the outside of the bend is increased (while the tensile stress in the steel on the inside of the bend is decreased). The maximum tension on any element of the casing must be less than the tensile strength of the casing.

Casing is pressure tested after cementing and this produces a force trying to pull the casing apart—a tensile force. The inside diameter of 9-5/8" 53.5 lb/ft casing is 8.535", which gives a cross sectional area of 57.2 square inches. A 3000 psi pressure test will impose an additional 172,000 lbs (78 tons) of tensile force on every joint in the casing, in addition to the weight. The top joint of casing in this case will therefore have to resist 707,000 lbs (320 tons) of tensile force (again neglecting buoyancy).

Compression. The major compressive forces come from buoyancy effects, which act at the bottom of the casing (and also act wherever there is a change in cross sectional area). If a single string of casing, 9-5/8" 53.5 lb/ft were used for a 10,000 ft. well, the buoyant upward force can be calculated. If the casing is filled with drilling fluid of the same density as the fluid outside it, then the buoyant upward force is simply the hydrostatic pressure on the bottom multiplied by the cross-sectional area of the pipe. This is illustrated in Figure 9-1. Once cement is placed outside the casing with mud on the inside, it's a little more complicated to calculate the net force, but calculating the weights and pressures involved follows the same principle.

The weight in air of the example casing string was 535,000 lbs. If a buoyancy force of 93,280 lbs is imposed then the weight in mud of this casing string is 441,720 lbs, measured at the top of the casing string.

The bottom of the casing will be in compression. The amount of compression will decrease moving up the string by the weight per foot. Therefore at 1000 ft. above the bottom, the compressive force will have reduced by (1000 x 53.5) 53,500 lbs. The neutral point for axial force, the point at which the casing is neither in compression nor in tension, can also be easily calculated if the upward buoyancy force and the weight/foot of the casing are known

$$\text{Height of neutral point above bottom} = \frac{\text{buoyancy force}}{\text{weight per foot}} = \frac{93,280}{53.5} = 1,744 \text{ ft.}$$

$$\text{Depth of neutral point} = 10,000 = 1,744 = 8,256 \text{ ft.}$$

Drilling Technology in Nontechnical Language

9-5/8" casing in 0.6 psi/ft mud

Hydrostatic pressure at 10,000' =
10,000 x 0.6 = 6000 psi

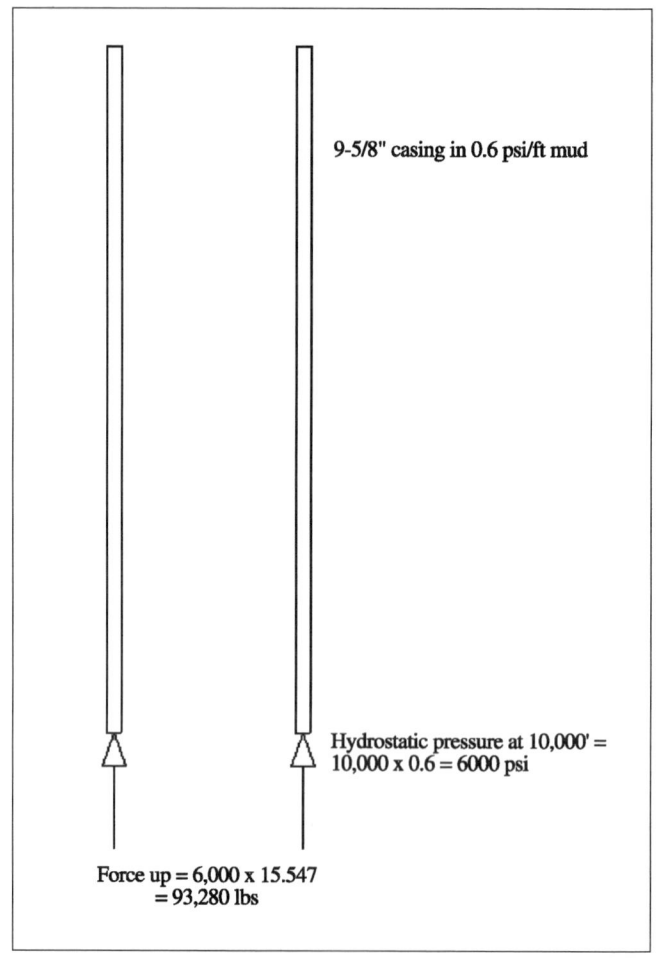

Force up = 6,000 x 15.547
= 93,280 lbs

Fig. 9-1 Buoyancy Effect on Casing

Burst. Casings must be able to withstand internal pressure. Internal pressure will come from downhole formation pressures, hydrostatic pressures, and pressure tests. If gas enters the well and is allowed to float upwards inside a closed well, the gas pressure cannot decrease as it moves up (because it can't expand). When the gas reaches the surface, it will still contain very high pressure that will then be added to the hydrostatic pressure all along the well. This can create extremely high pressures and it must be considered for the production casing.

Collapse. The opposite of internal pressure is where the pressure outside of the casing is higher than the pressure from the fluids inside the casing. It is possible for casing to be squashed as flat as a ribbon by external pressure.

Cemented casing is much harder to collapse (takes a much higher pressure) than uncemented casing.

Flowing salt was discussed in chapter 1. Thick salt deposits flow under pressure from the rocks above. Thus salt acts like a hydraulic fluid in these conditions and can impose very high collapsible pressures on a string of casing. Two factors are important when cementing in flowing salts—the casing has to be very strong (thick wall and/or high strength steel) and the cement around the casing must form a complete sheath. If there are unfilled areas without cement, the salt can flow into that area and place a very high point loading (as opposed to an even loading) on the casing. No casing can resist a high point loading.

Driving forces. Conductor pipes are sometimes driven (hammered in place) by a pile driver into the ground to allow the well to be spudded with a closed circulating system. Conductor pipe is thick walled (often 1" or greater wall thickness), so the the pipe is strong enough to drive. The connections must be selected to be suitable to transmit the heavy shock loads of driving.

Buckling. Buckling is a stability failure of a long tube. It is only a potential problem when a long string of casing is uncemented in a vertical well and when either the temperature or the internal pressure will increase significantly after the cement has hardened (Fig. 9-2).

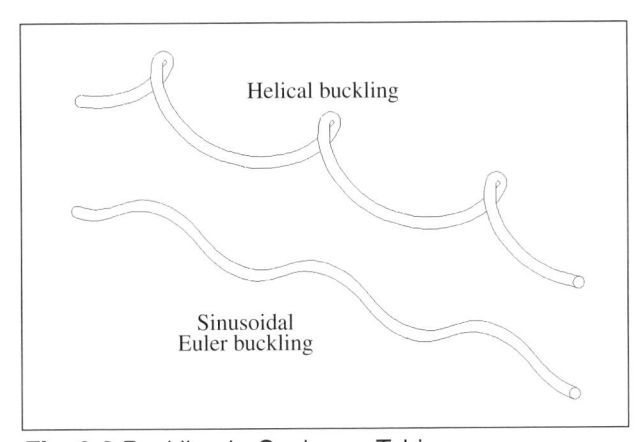

Fig. 9-2 Buckling in Casing or Tubing

Many people assume that buckling is only possible when the tube is in axial compression, but this is not true. The internal and external pressures also affect buckling. Casing suspended in the well is in compression at the bottom due to buoyancy forces, but the casing does not buckle here because the internal and external pressures (that are equal when the same fluid is in the annulus and inside the casing) act together to stabilize the casing. The neutral point for buckling in this case is actually at the bottom of the casing, while the neutral point for tension/compression (or axial force) for the example casing is 1,744' higher, as was calculated earlier.

Above the neutral point for buckling, the casing cannot buckle. Below the neutral point, the casing may buckle depending on the resistance to the buckling of the casing itself. Support is also important; if the casing is cemented (100% support) then it won't buckle. If it is well centralized inside an in-gauge hole it won't buckle. If the hole is at an angle, the support given by the borehole wall reduces buckling and, in fact, once hole inclination exceeds about 20° it would be quite hard to make the casing buckle.

Temperature. When a steel casing gets hot, it expands. Where the casing is cemented this does not cause any problems but the uncemented pipe between the top of the cement and the surface wellhead may buckle as expansion takes place. This can be compensated for by stretching the casing before setting it in the wellhead.

Steel also loses strength as temperature increases. In a deep, hot well this loss of strength can be significant. At a temperature of 200° C, steel will have lost 19% of its strength. This has to be accounted for when designing casings for high temperature wells.

Combined axial and internal forces. If a steel tube is in tension, it has an increased resistance to burst and a decreased resistance to collapse. Conversely, if a steel tube is in compression it has a decreased resistance to burst and an increased resistance to collapse. This is known as the *biaxial effect*. The industry standard design tables for casing (published by the American Petroleum Institute, (API)) which give strengths in tension, collapse, and burst also include information on biaxial effects. Generally when designing casing, the increased burst strength due to tension is not allowed for (increasing the safety margin), but decreased collapse resistance due to tension will be accounted for where it is relevant.

Corrosion. When acidic gases such as hydrogen sulfide (H_2S) or carbon dioxide (CO_2) are present with water, steel components can become seriously corroded. This is worse with high temperatures; corrosion rates roughly double for every 32°C increase in temperature. It is also worse with higher pressures and with higher concentrations of corrosive agents. If H_2S, CO_2 and water are all present, then corrosion resistance design will require a special study to determine the most cost effective solution.

Special steels containing nickel or chromium can be used but these are much more expensive than plain vanilla carbon steels. These materials are also softer than carbon steel and the connections are easily damaged while screwing joints together (if spun too quickly or if not exactly aligned while turning).

Hydrogen sulfide is a particular problem for designing casings. Hydrogen can enter the steel crystalline structure and causes "hydrogen embrittlement". The steel can break at well-below expected failure loads without warning if this has happened. The problem gets worse with higher strength steels and at lower temperatures. Lower temperatures are found at the top of the casing string, where the tensile load is highest. If H_2S is encountered in the well, it is very important that the correct type of steel (the grade) is selected so as to avoid hydrogen embrittlement and failure at the exposure temperature of the steel.

Connections. Most failures in casing (around 90%) occur at the connection—the part that screws two joints together. This should not be surprising; an extruded steel pipe is pretty strong whereas a pipe that has a thread cut on it must have decreased strengths against some, if not all, forces.

Particular strain is placed on a connection where the casing is placed in a curved section of wellbore. High tension combined with bending and internal forces, places great strength requirements on the connection threads.

Casing and tubing connections might not seal (just contain a thread) or there may be an "O" ring or elastomer seals incorporated. For gas well service, metal to metal seals are normally called for. These work by placing the two metal seal areas in the pin and box in contact just before the connection is fully screwed together. As the joint is torqued, energy is placed into the seals that cause some elastic deformation and creates a tight seal between the metal faces.

Role of the Cement Outside Casing

Casing is cemented in the annulus, either the complete annulus or sometimes just the lower part. The cement is designed to meet a variety of needs:

1. Physically support the weight of the casing
2. Prevent fluids from migrating upwards inside the cement, between the cement/formation, or casing/cement interfaces
3. Protect the casing against corrosion
4. Protect the casing against mobile formations
5. Allow the production casing to be perforated without the cement shattering under the shockwave

Mud Removal

One of the most difficult aspects of cementing is to remove all of the drilling fluid from the annulus so that it can be replaced by cement. In an in-gauge hole with well centralized casing, the chances of achieving full mud removal are good. In an enlarged hole with the casing not well centralized, the chances are very poor. Great care is needed in designing and executing the job.

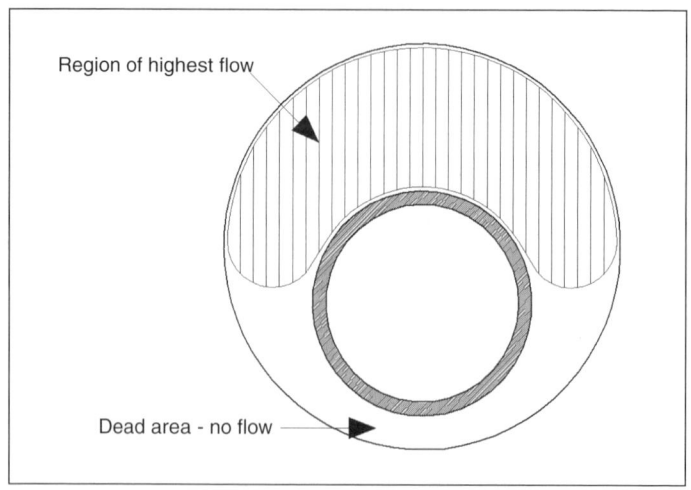

Fig. 9-3 Cement Flow in an Eccentric Annulus

Chapter 9 • Casing and Cementing

Figure 9-3 above shows what happens when the casing is not centralized. The cement will preferentially flow to the largest cross sectional area, leaving mud in the narrow part. The closer the casing is positioned towards the wall, the harder it will be to remove this mud. Several actions can be taken to maximize the chances of full mud removal

1. Drill a stable, in-gauge hole.
2. Tailor the mud properties before running casing so that the mud is as thin as possible.
3. Move the casing (either rotate it or reciprocate it) while pumping cement around. This causes the pipe to move around in the well so that there isn't a single area of no flow. Pipe movement is proven to be very beneficial in achieving mud removal.
4. Pump thin spacers ahead of the cement. This separates the mud from the cement. The spacers can be tailored to help chemically clean filter cake from the wall and casing, leaving the surfaces water-wet and ready to bond to the cement.
5. Use centralizers to keep the casing in the center of the hole.

It is important that the casing is well "centralized" and to ensure this, a tool is fastened to the outside of the casing that pushes the casing off the wall. These tools are called *centralizers*. A centralizer is made of strips of spring steel, held together at the top and bottom with a steel ring that hinges open so that it can be placed on the pipe. A centralizer is shown in Figure 9-4.

Fig. 9-4 Casing Centralizer

Cement

Cementitious material consists of a powder that undergoes chemical reactions when mixed with water. The end result of these reactions is a hard, stone-like material. Cements have been used for centuries (the ancient Egyptians, Greeks, and Romans all used cementitious materials). Modern cement development started with a British patent granted to Joseph Aspdin in 1824 that defined the process used to manufacture cement for building a lighthouse. This cement was called *Portland* cement because when hard, it looked like stone from the Isle of Portland, used for building.

Portland cement has four principal components—tricalcium silicate (Ca_3S—about 70%), dicalcium silicate (Ca_2S—not more than 20%), tricalcium aluminate (Ca_3Al), and tetracalcium aluminoferrite (Ca_4AlF). These chemical compounds are created by mixing raw materials together and firing in a kiln at high temperatures (up to 1500° C).

The raw materials are lime, silica, alumina, and iron oxide. Before firing they are finely ground and mixed in the correct proportions. After firing, the raw materials have been converted to a material called *clinker*. After cooling, some gypsum ($CaSO_4 \cdot 2H_2O$) is added (3%-5%) and the mixture is crushed to a powder. This powder is Portland cement.

The chemistry of setting cement is quite complex and several stages can be identified. However, in general terms, when water is added the components form hydrated compounds. Crystals are formed that grow and become interlinked with one another. Also in the early stages, materials are dissolved in the water and later on are precipitated as solids. Some water is left trapped in the spaces between the crystals—cement is porous, but should be impermeable because the passages connecting the pores are small enough so that movement of water is stopped.

The reaction of hydrating Portland cement is exothermic—it generates heat. This can be a problem in arctic areas when drilling through the permafrost. Special cements are then used that do not freeze and have a low heat of reaction while developing sufficient compressive strength to meet the requirements of the set cement.

The API established standard specifications for oil well cements. This defined eight different cements, which were classified according to the depths and temperatures at which they could be used. These classes were designated A through H. The specifications state the chemical and physical attributes of the cement. Over time, one class has proven to be the most useful when its properties are chemically modified during mixing. This is

API Class G cement. It is universally available around the world and there is a vast amount of experience in using it.

An amount of Class G cement powder will require a certain volume of water to hydrate it and make it pumpable. 22% water by weight of cement (BWOC) is needed to completely hydrate the cement, but this would not make a pumpable slurry. 44% water BWOC gives a pumpable slurry and the "extra" 22% water is held within the set cement matrix. Excess water (above 44% BWOC) will be left as "free water" after the slurry sets. The point where the correct amount of water is used to make a pumpable slurry with no free water is known as "neat" cement. For API Class G cement, the water requirement is 4.96 U.S. gallons for each 94 lb sack and the resulting slurry weight is 15.8 lbs per gallon. Free water for normal slurries should be no more than 0.5% of the slurry volume, 0% for a slurry designed for high angle or horizontal wells.

Cement Design

Density

Neat Class G cement can be modified to suit particular requirements of the well. The most important slurry property is density. As noted, neat cement slurry weighs 15.8 lbs per gallon (ppg), which equals a density gradient of 0.822 psi/ft. Normally, casings are cemented with two different slurry densities—a light "lead" slurry and a neat "tail" slurry (Fig. 9-5). This is done for two main reasons:

1. **Hydrostatic pressure.** A long column of neat cement slurry might cause formations downhole to break down due to the high pressure
2. **Cost.** The light slurry does not require as much cement powder and additives and is cheaper

As noted, if more water than 44% BWOC is used, free water will be left when the slurry sets. Free water must be minimized because this can form channels through the cement that will later allow fluids to flow. Cement is made lighter (lighter cement is called *extended* cement) by adding more water. To soak up the excess water, clay is added in sufficient quantity so that free water is not left after setting. The clay used is bentonite and when it is used to lighten slurry, it is called an *extender*. Standard cement design tables give the quantities of cement, water, and bentonite needed to mix slurry of various densities. For instance, to mix one U.S. gallon of slurry using API Class G cement, the following mix is required at different weights:

Slurry weight	Cement required	Water required	Bentonite required
14.2 ppg	8.2 lbs	0.66 U.S. gallons	0.33 lbs
13.2 ppg	6.7 lbs	0.73 U.S. gallons	0.54 lbs
12.6 ppg	5.6 lbs	0.76 U.S. gallons	0.67 lbs

Table 9-1 Slurry Weight Requirements

Other materials may also be used as extenders. Hollow glass or ceramic microspheres can be added to neat cement, as can materials with low specific gravity, such as powdered coal, crushed volcanic glass, or gilsonite[1]. Cement can also be made into foam by mixing with nitrogen, which can give very light slurries for weak zones that cannot handle much hydrostatic pressure. Slurries down to 7 ppg (0.364 psi/ft) can be created using foam.

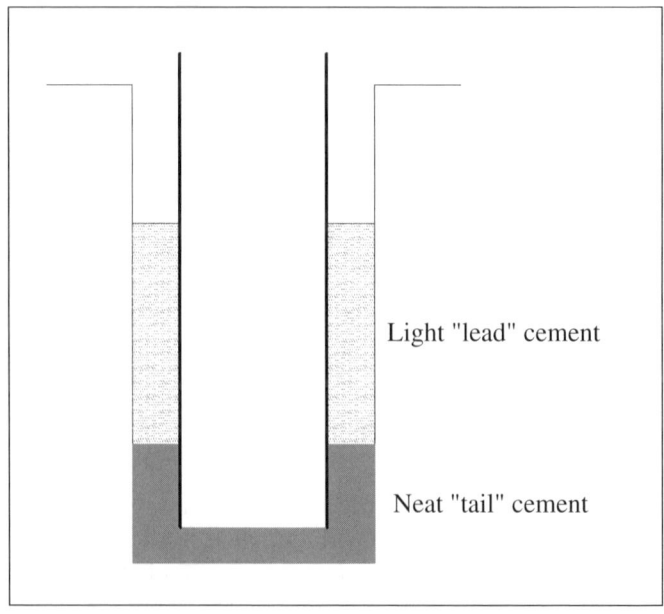

Fig. 9-5 Cement Slurries Used Outside Casing

It may also be necessary to mix very heavy slurries that are denser than neat cement. In this case, heavy materials such as barite are added to the slurry.

The lower limit of cement slurry density is dictated by the requirement to always maintain a hydrostatic overbalance on pore fluid pressures while pumping cement around and into place. The upper limit is dictated by the strengths of downhole formations.

Thickening time

The thickening time of a slurry is tested as part of the slurry design. Procedures for testing are given by API Specification 10, which measures the time it takes for the slurry to reach a consistency of 100 Bearden units (Bc) at downhole temperature and pressure. The Bc is a dimensionless value that cannot be directly converted to oilfield viscosity units such as poises. While the test measures the time to reach 100 Bc, it is generally accepted that the limit of pumpability is reached at 70 Bc. The test lab can be asked to provide both values. The thickening time generally should be enough to displace the cement and circulate it back out if problems occur.

Compressive strength

To perform the various functions, cement is designed to provide a high compressive strength. A sample of the cement slurry is made up and placed into a 2" cubic mold. Once the cement is set, this cube is taken out of the mold and placed into a hydraulic press, where it is squashed until it breaks. The pressure required to break this cube is measured. 500 psi is considered to be a good compressive strength to support the casing, 2000 psi is considered the minimum for cement that will be perforated.

Temperature rating

The thickening time and compressive strength buildup are dependent on well temperature. Higher temperature gives faster setting and faster strength buildup. The slurry must have sufficient pumpable time to complete the job, with a safety margin in case of problems. Also the thickening time should not be so long that rig operations are unnecessarily delayed while waiting for cement to set. The thickening time is determined in the laboratory using samples of cement, chemicals, and mix water sent in from the rig.

Drilling Technology in Nontechnical Language

Accelerators or retarders (chemical additives) can be used to lessen or lengthen the pumpable time and will similarly affect the rate of compressive strength buildup. Under static (non-pumping) conditions, the well will have a temperature gradient as the formations get hotter with depth—normal temperature gradient in sedimentary basins is about 1.4° C per 100' of vertical depth. Circulating will decrease the local temperature around the wellbore. Thus at any particular depth, two working temperatures will be relevant to cementing operations—circulating and static.

A temperature log run on wireline some hours after finishing circulating will give the static temperature at the bottom of the well (BHST). The circulating temperature at depth (BHCT) can be calculated by reference to API Specification 10, which contains temperature tables. It is also possible to measure this temperature directly during circulating with small thermosensitive probes. Of these two temperatures, BHST is relevant to investigating cement stability and compressive strength development with time. Bottom hole circulating temperature is used when calculating pumpable time.

As a rule of thumb, the static temperature at the depth of the top of the cement should not be less than the BHCT used in slurry design. If it is significantly less, it may take an unacceptable length of time to cure; in this case, extra testing should be done at the actual TOC static temperature to see if the cement characteristics are still acceptable.

For deep, hot wells (BHST > 110° C [230° F]) the long term stability of Portland cement requires the addition of silica flour, usually 35% BWOC. If silica flour is not added, the strength of the set cement will slowly decrease with time.

Rheology

Rheology was covered in some depth in chapter 5 when discussing drilling fluids. The cement slurry rheology is very important because this will affect downhole pressures while pumping cement around the casing and up the annulus. It will also affect mud displacement, mixability, pumpability, and free fall[2] of the slurry down the casing.

Cement slurry rheology is very complex and depends on many factors, such as the

- Ratio of solids (cement powder, bentonite, etc) to water
- Sizes and shapes of the solids present in the slurry

- How much energy was used to mix the slurry (affects the distribution of particles and the speed of chemical reactions)
- Flow regime (laminar, turbulent, or transitional)
- Time—the cement rheology continuously changes as chemical reactions take place
- Temperature and pressure—the rheology changes as the cement moves down the well

Even though tremendous efforts have been made by the industry to completely characterize and explain cement slurry rheology, this work is not yet complete. As with mud rheology, the best model currently available to describe most cement slurry rheologies in the field is the Herschel-Buckley model.

Chemical additives

All characteristics of the cement can be modified by adding chemicals to the slurry. These are called *additives*. Some have already been mentioned, such as bentonite or powdered coal as extenders, barite as a weighting additive, retarders (to slow down the setting speed of the slurry), and accelerators (which make the slurry set faster). Other additives available include:

- Defoamers—prevents the slurry from foaming while mixing
- Dispersants—help to distribute the solid particles present in the slurry
- Fluid loss—controls loss of filtrate into permeable formations
- Lost circulation material—inert solid materials used to plug off pore spaces at the formation face to prevent the loss of slurry to the formation

The use of additives allows one cement type (API Class G) to be used for many different wells and in different applications.

Cementing casing in massive salt formations

When cementing in massive salts, the cement forms an essential part of the casing string integrity. Inadequate cement will make shearing, distortion, or failure of the casing possible as the salt moves. The potential failure modes include:

1. Point loading of the casing due to uneven salt closure. The casing can collapse with much less force than would be the case for even loading.
2. Collapse due to overburden pressure being transmitted horizontally by mobile salt.
3. Shearing of the casing due to directional salt flow.
4. Corrosion of the casing, particularly if magnesium salts are present.
5. Long term degradation of the cement sheath by ionic diffusion into the cement, if not salt saturated. If the cement sheath degrades, uneven loading may occur leading to eventual collapse.

The contributing factors to these failures include:

- Salt creep causing hole closure. This occurs faster in bigger holes and is proportional to hole diameter; a 16" hole will reduce in diameter twice as fast as an 8" hole.
- Lowered hydrostatic pressures will increase the rate of creep.
- Salt flow due to directional field stresses.
- Leaching by mud and cement leading to overgauge hole and slurry chemistry alteration.
- Ionic diffusion of salts into a non-saturated slurry after setting. Magnesium is particularly detrimental.

The essential objectives are to cement throughout the whole salt body interval, to ensure that good cement completely fills the annulus and to prevent long term degradation due to ion diffusion. There are several things that can be specified in the drilling program to maximize the chance of success:

- **Use a salt saturated slurry.** If the slurry is unsaturated at downhole temperature, substantial quantities of salt can be leached out by the slurry. This will give an overgauge hole and will significantly affect thickening time, rheology, and compressive strength. Supersaturating a slurry may involve heating the mix water to dissolve more salt. At these saturations, special additives (especially dispersants and fluid loss) are needed. Saturated KCl slurries give higher compressive strengths faster than saturated NaCl slurries. Setting time is important (see the next point). Salt saturated slurries can cause problems against other formations. If exposed long term exposure to unsaturated formation water allows osmotic forces to leach salt out of the cement slurry, which can lead to cement failure. This may or may not be a problem, depending on what formations are exposed and where they are.

- **Use fast setting times.** Once cement starts to set, it holds hydrostatic pressure from above. This reduces pressure on the salt. As pressure is lost, the salt creep rate will increase substantially. With long setting times, the salt could creep in enough to touch the casing. As salt does not creep uniformly, the resulting point loading on the casing will quickly collapse or deform it. Even the strongest casing cannot resist such point loadings.

- **Use suitable drilling fluids to minimize leaching the salt.** Large washouts will lead to the normal problems of mud removal and this will lead to an incomplete cement sheath. However using oil- or salt- saturated water mud can give problems as the hole will close in while drilling.

- **Increased mud densities.** Reduce the rate of creep.

Low salt slurries have been used successfully in the Gulf of Mexico and other areas. These aim to give fast development of high compressive strengths. These slurries will avoid problems against other formations due to osmosis as mentioned above. However, washouts are still likely and long term ion diffusion may later become a problem.

Clearly, cementing against massive salts is a complex problem if the well is to meet its long term objectives. The success of this cement job starts when drilling through the salt (minimizing leached washouts). Good planning, expert involvement and attention to every detail including slurry and

Drilling Technology in Nontechnical Language

spacer design, rig equipment, downhole casing configuration and cement job supervision/quality control is vital.

Running and Cementing Casing

After drilling and logging a hole section, casing is lowered into the well. It is cemented in place by pumping cement down the inside, where it exits at the bottom of the casing and comes back up the annulus.

Casing usually comes in lengths of around 40'. On the bottom joint of casing a special valve is screwed, called a *float shoe*. This is made of cement with a plastic valve in the middle, so that it can be drilled with normal drill bits. The valve allows fluids to be pumped down through it, but does not allow fluids to flow upward. This keeps the cement slurry in place once pumping stops. Without the float valve, the higher hydrostatic pressure in the annulus would force cement to come back up inside the casing once pressure is released at the surface after cementing. Pressure in the annulus is higher because cement slurry in the annulus is denser than the mud inside the casing.

Usually two joints of casing are run above the float shoe. Then another float valve is screwed in place, called a *float collar*. One purpose of this is that if the valve in the float shoe fails, there is a backup valve in place to stop flow back into the casing.

With the float shoe and float collar made up, more joints of casing are screwed on top of the string and lowered in to the well. Centralizers are attached as are needed due to the hole deviation and condition. Once the casing is all in the well, a hanger can be made that is used to suspend the casing inside the wellhead. This was explained in chapter 3.

With the casing landed in the wellhead, it is now necessary to prepare for cementing. A special container is screwed on top of the casing; this holds two plugs with rubber fins that fit inside the casing (Fig. 9-6).

The bottom plug is hollow and has a thin rubber diaphragm at the top. When ready to start pumping cement, the lower releasing pin is drawn out and cement is directed to the inlet pipe between the two plugs. The bottom plug travels ahead of the cement, separating it from the mud below to minimize contamination of the cement with mud. The fins on the bottom plug, as well as sealing against the inside of the casing, wipe the thin film of mud off the inside of the casing, again, to reduce contamination.

Once the correct volume of cement is pumped, the upper releasing pin is withdrawn and mud is pumped into the upper inlet. Now there is a

Chapter 9 • Casing and Cementing

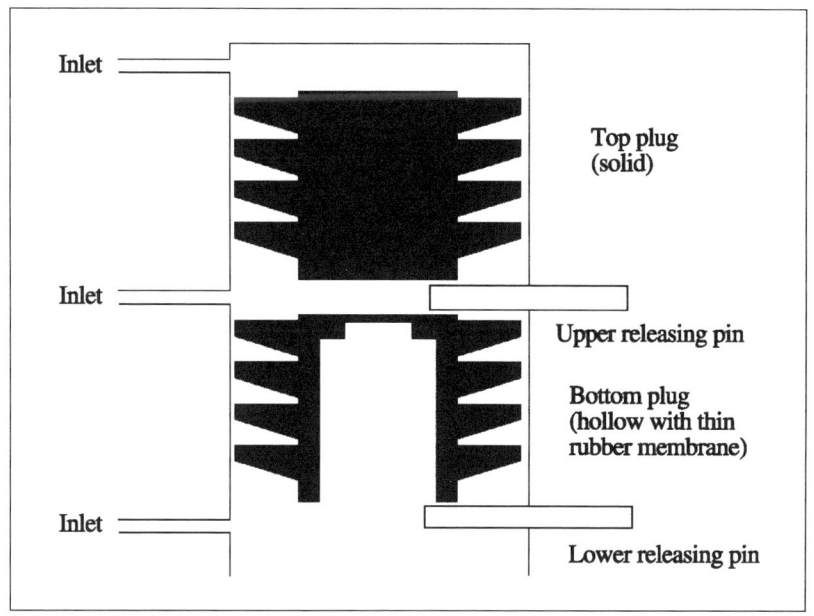

Fig. 9-6 Cement Plug Container

column of cement moving down the casing, with the bottom plug below and the top plug above.

Eventually the bottom plug hits the float collar. It can move no further down. At this point the rubber membrane ruptures to allow cement to flow through the plug. Cement now flows out of the float shoe and up around the casing.

The pumps are slowed down as the top plug approaches the float collar. Once the top plug lands on top of the bottom plug, it forms a seal. The surface pumping pressure increases and this shows that the displacement is complete.

Normally at this stage, the casing is pressure tested. If for some reason the top plug did not land on the float collar after pumping the correct volume, pumping is stopped and the casing will have to be pressure tested once the cement has set. This involves a small risk of breaking the bond between the casing and the cement. The casing will expand a little under the test pressure; when pressure is removed, it will contract again. However, the cement is not elastic like steel and instead of moving back with the casing, it's possible that the bond will break. This creates a very small annulus between casing and cement—a *microannulus*.

Drilling Technology in Nontechnical Language

Cementing Surface Casings

Surface casing is a large diameter, so that several more strings of casing can fit inside it and still have a big enough hole through the reservoir to be able to log it and produce from it. Large diameter casing is difficult to handle and to run, especially in any significant wind. A 40 ft. joint of pipe 20" in diameter and weighing around 4,000 lbs will be tricky to manipulate in bad weather conditions. If the top of the casing is slightly off center, the casing connection thread is likely to become cross-threaded, where the threads on the male end (bottom of the joint of casing) do not quite line up with the threads on the female end (top of the previous joint). The connection can be rotated but the threads become damaged and will not fully make up, so it has to be unscrewed again. The threads might be so badly damaged that both of the casing joints have to be taken off again and replaced. It's no fun running big casings!

The casing is run with a special float shoe on the bottom, called a *stab-in float shoe*.

When the casing shoe is at the correct depth, the top of the casing is held in the rotary table. The rotary table suspends the weight of the casing. Then drillpipe is run inside the casing, with a special seal assembly attached to the bottom of the drillpipe. This seal assembly locates in a hole on the top of the stab-in float shoe (it "stabs in"), so that a seal is formed between the drillpipe and the casing shoe. Now, mud and cement can be pumped down the drillpipe, out of the float shoe and up the annulus. When drillpipe is used in this way, the drillpipe is referred to as a "stinger[3]" (Fig. 9-7).

Surface casing is always cemented from the bottom to the surface—the complete annulus. The volume of the hole is not accurately known, so the amount of cement needed to fill the annulus is also unknown.

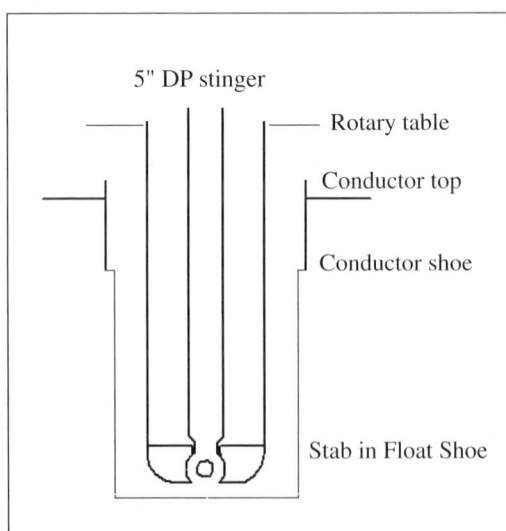

Fig. 9-7 Diagram Showing How the Stinger Seals in the Casing Float Shoe

210

Chapter 9 • Casing and Cementing

With a drillpipe stinger, cement can be pumped down the stinger and up the annulus until cement is seen coming out of the annulus at the surface. Once cement returns are seen at the surface, mud is pumped to push the small volume of cement left in the stinger to the float shoe.

If cement was pumped down the casing without a stinger, at the point that cement starts coming out of the annulus there would still be a huge volume of cement inside the casing. This cement has to be removed from the inside of the casing by pumping down mud behind it; that means that a lot of cement is wasted (with surface disposal problems for such a lot of cement). Also, if there is a problem with the cement, it might harden inside the casing and have to be drilled out. This would take a long time and it would also be very embarrassing!

Once the cement is displaced into the annulus and the drillpipe stinger is full of mud, the drillpipe can be pulled out of the hole.

Cement Evaluation Behind Casing

It is possible to run wireline logs to evaluate the condition of the cement behind the casing. This is done by sending sound waves from a wireline tool and listening for the returning sound wave a short distance away. As the sound travels up the cement, the frequency and amplitude are affected by the quality of the cement and of the cement-casing and cement-formation bonds.

A basic cement evaluation log will give an indication of the bonds, but not much else. More sophisticated tools can detect the presence of a microannulus, channels in the cement, gas bubbles in the cement, and much more. But, of course, this tool costs a lot more to use.

Most cement jobs are logged, even if only with the basic tools. In critical cases, especially where problems are suspected, a better analysis of the actual problem is required before any remedial work can be considered.

Other Cement Jobs

Secondary cementing

Cement is pretty versatile stuff, thanks to additives, and this gives many possibilities for using cement to solve a variety of problems. If a channel is detected in the cement behind the casing in a critical cement job, it may be possible to perforate the casing and force cement into the channel under pressure. Repairing a bad primary cement job (primary = the first time that the casing was cemented) is called *secondary cementing*. Unfortunately, secondary cementing is quite tricky and has a low success rate.

The general procedure is that holes are made in the casing using a perforating gun. Drillpipe is run in the hole down to the depth of the perforations. Cement is pumped down the drill string where it exits around the perforations. Pressure is then applied to try to force cement to enter the perforations. If lady luck is smiling, maybe the cement will plug the channel and the perforations. Forcing cement in under pressure is called *squeeze cementing*.

A better chance of success comes if the top of the cement is just too low. It is then possible to perforate above the top of the cement. Circulation can be established, hopefully, by opening the annulus outside the casing and pumping fluid into the casing. If circulation is possible, cement can be placed by circulation rather than by squeezing. It is also possible that bits of formation (cuttings and cavings) have settled around the casing and this can prevent circulation.

This technique is also used when wells have to finally be abandoned, or if for any reason (i.e., government regulations) some or all of the annuli require to be full of cement.

Curing lost circulation

Lost circulation can occur from a variety of causes. This will be discussed in more detail in chapter 13.

If serious circulation loss occurs, cement can be used to cure it. The difficulty is in getting the cement in the right place and keeping it there while it sets.

Cement plugs

A cement plug is a column of cement that is set at some point in the well. Cement plugs serve a variety of purposes.

During well abandonment, cement plugs are set at various points inside the casing to prevent downhole fluids from reaching the surface in the future. Cement plugs are also used to suspend a well (temporarily abandon it), for instance, if the well will be completed later on by another rig. In this case, the well will be re-entered and the cement plugs drilled out before continuing with completion operations.

If for some reason the lower part of the well is to be re-drilled along a separate path, a cement plug may be used to:

1. Abandon the lowest section of the original wellbore
2. Allow the drill bit to depart from the original bore and drill a new one

The main criteria for this kickoff cement plug is that the set cement must have a higher compressive strength than the surrounding rock, or else the bit will drill back along the original hole.

Sometimes formations exposed in the wellbore can be very unstable. They may be fractured (either naturally or as a result of the drilling operation) or they may react with the drilling fluid in some way. This results in material becoming detached from the wellbore wall and falling in to the well. The hole enlarges. This might be cured by setting cement across it to fill the enlarged hole and isolate the troublesome formation. The cement is then drilled through to form a cement lined hole.

Cement plugs are set by running tubing into the well and pumping cement down the tubing into the well. The tubing is then withdrawn, leaving behind a column of cement. Setting a successful cement plug the first time is more difficult than achieving a good cement job outside the casing; the annular capacity is larger (because tubing or drillpipe is smaller than casing) and annular velocities are lower, so complete removal of the mud is harder to achieve. The slurry must be designed so that once in place it doesn't move as the tubing is removed, or afterwards while it is still fluid. Contamination of the cement by mud in the well is a real problem, which can be solved by using a special tool on the bottom of the tubing that directs the flow of cement outwards and upwards, instead of straight down as would occur with just plain ended tubing. Cement plugs require good planning and careful execution to meet the objectives without having to be repeated.

Chapter Summary

In this chapter, the major elements of casing design were discussed. The criteria by which a casing design is judged is whether it meets the requirements of the well design at the lowest cost. Casing costs generally form around 10% of the total cost of the well, so it's a major element in well design.

Cementing of casings was described, with the two techniques (plug cementing and stinger cementing) being covered. The most important cement design parameters were touched upon.

Cement plugs for various purposes were briefly described.

It can be seen that good casing, cement design, and execution are critical to the safe and cost effective drilling and producing of oil and gas wells. Time (and therefore money) spent on design in these areas is never wasted. Engineers involved in well design should receive training to keep them abreast of the latest developments and should have access to the best design tools available (computer programs, reference information, and expert advice for difficult cases). They should also have the time to complete a thorough review and design job. Unfortunately, in many cases, not all of these elements are present.

Glossary

[1] **Gilsonite.** A naturally occurring black solid with a low specific gravity of 1.07. It is asphaltic (predominantly a mixture of heavy hydrocarbons forming a solid material) and starts to become soft above 240° F and melts at 385° F. For this reason, gilsonite is not recommended as a cement extender in wells where the bottom hole temperature exceeds 300° F.

[2] **Free fall.** When cement slurry is pumped into the casing, the total hydrostatic pressure of the fluids inside the casing is greater than that of the fluid in the annulus, because cement slurry is heavier than mud. This leads to a condition where the cement will continue to fall down the casing, even if pumping is stopped. With a large cement job, it is possible for this U-tubing effect to cause the cement to fall faster than the pumps can fill the casing behind it, causing a partial vacuum inside the casing. This condition is called *free fall*.

[3] **Stinger.** A stinger is a small diameter pipe that is used for accurately placing fluids down the well.

Chapter 10

Evaluation

Chapter Overview

The term "Evaluation" covers several different techniques that, together, allow different disciplines within the team (Drilling, Geology, Production, Reservoir Engineers, etc.) to understand the subsurface conditions. Typical information obtained will describe lithology, pore fluids present, wellbore condition, rock structures (bedding planes and faults), and detailed reservoir characteristics (static and dynamic).

This chapter will describe the different types of techniques available as well as the type and quality of information possible from each.

Evaluation Techniques

The techniques available can be divided in to the following major categories

1. Physical sampling at surface (examining drilled cuttings and returned drilling fluid, also recording data while drilling)

2. Physical sampling downhole (taking cores of rock or samples of fluids and recovering them to surface)

3. Electrical logging (using downhole instruments to measure physical attributes of the formations, casing, and cement)

4. Production testing (flowing hydrocarbons from the reservoir while measuring pressure)

Drilling Technology in Nontechnical Language

Physical Sampling At Surface

In the early days of drilling, the main source of information on the formations being drilled through came from examining the drilled cuttings returned to the surface. The depth that the sample came from can be estimated (but not known precisely) by recording the time that the sample appeared at surface and by knowing the time taken to circulate a cutting from the bottom. Subtracting the transit time from the time on surface allows the wellsite geologist to work out the bit depth at the time that the cutting was generated, and so gain an idea of the depth from which the sample came.

A solid particle will fall through the drilling fluid in the annulus at a speed that depends on it's size, shape, and density relative to the density of the fluid. This downward speed is called the "slip velocity". During circulation, the upward speed of the fluid in the annulus is called the "annular velocity". The net speed at which a particle will move up the annulus is calculated by subtracting the slip velocity from the annular velocity. As drilled cuttings from one formation are generated in a variety of sizes and shapes, it follows that their slip velocities will differ, too. Samples from one specific depth will arrive at the surface over a period of time and not all at the same time. For this reason, the actual depth of a particular sample is unlikely to be precisely known.

Another source of inaccuracy in sample depth determination is that some cuttings may settle at the side of the hole (in a deviated well) or in an enlarged section of the hole. Later, pipe movement or an increase in pump speed might disturb these beds of cuttings and allow them to continue upwards. In this case, samples from much higher in the well might suddenly appear at surface, mixed in with the newer (deeper) cuttings. Plenty of scope for confusion!

When the bit drills from one formation into a distinctly different formation, there will almost always be a detectable change in the rate of penetration or drilling torque, if the WOB and RPM are kept the same. The depth at which this ROP or torque change takes place is used to adjust the sample depths for greater accuracy. For instance, if drilling from shale into salt, there will usually be a sudden increase in the rate of penetration. The first samples of salt arriving at the surface should have come from the very top of the salt formation, though mixed in with this sample will be particles of shale still returning up the annulus. The depth of this first salt sample can be assumed to be the depth where the ROP increase was noted.

Chapter 10 • Evaluation

Drilling mud circulated around the well can also provide valuable information. Pore fluids or gas will enter the mud, even if a kick has not occurred, simply because pore fluids will be released from the drilled cuttings coming up the annulus. This might then cause a detectable change in mud chemistry, for instance, by increasing the salinity of the mud.

Just a few rocks and minerals make up most of the Earth's crust. All of these are readily identified by simple tests. It is important for the geologist to identify these rocks and minerals in the field without elaborate equipment. Minerals generally occur as small grains making up the rocks. The main attributes that the geologist uses to identify a rock are described next.

Color. Color is the first property that may be noticed in minerals. Some transparent minerals are shaded different colors by slight impurities such as iron or titanium. Certain types of quartz are good examples of this.

Luster. Luster is the appearance of light reflected from the surface of a mineral. Two typical lusters are metallic and non-metallic. Non-metallic lusters have self-descriptive names such as greasy, glassy, silky, and earthy.

Transparency/translucency. A few minerals are transparent in thin sheets, while others are translucent (they transmit light but not an image). The majority are opaque (do not transmit light).

Crystals shapes and form. The form that a mineral crystal takes can also be diagnostic. Some minerals have characteristic crystals, such as cubes or pyramids, whereas other minerals have no crystal form.

Cleavage. The tendency for some minerals to break along planar surfaces is called cleavage. Three aspects of mineral cleavage are

- the number of cleavage surfaces of different directions
- the quality of the surfaces (e.g. poor, excellent, pitted, etc.)
- the angle between the surfaces

Hardness. The hardness of a mineral is quantified by Moh's scale, which ranges from one to ten. A mineral that is higher on the scale can scratch a mineral that is lower on the scale. Moh's scale is in order of hardness.

1. Talc
2. Gypsum
3. Calcite
4. Fluorite
5. Affatite
6. Orthoclase
7. Quartz
8. Topaz
9. Corundum
10. Diamond

Saltiness. Tasting can sometimes identify a mineral. (Not recommended with oil based mud!)

Acid test. A very important test is the application of cold, dilute hydrochloric acid to a sample. Calcite and the sedimentary rock limestone are made up of predominantly calcite ($CaCO_3$) grains and are the only ones that will bubble with dilute acid. Dolomite bubbles slightly–increased bubbling will be noted if heat is applied.

Swelling properties. Reaction of clays (hydration) to water or dilute acid.

Sorting. Distribution and estimation of grain size.

Roundness. Describes variety of shape from very angular to well rounded.

Sphericity. Describes shape from very elongated to very spherical. High Sphericity (spherical), Low Sphericity (elongated).

Cementation. The degree of cementation and the mineral type. May be *well cemented, good cementation,* or *poor cementation.*

Specific gravity (weight). Specific Gravity is the relative weight of a mineral compared to the weight of an equal volume of pure water. The average specific gravity (SG) of a rock or mineral would be about 2.5. Metallic ore minerals generally have specific gravities above 3.5. Minerals can be readily recognized from their SG.

Fractures. Some rocks are naturally fractured. Often these fractures will fill up with minerals and can be seen as veins of a different color within the rock.

Fluorescence. Crude oils will fluoresce under ultraviolet light (they will glow giving off light of a different wavelength to the source light). If a cutting from an oil bearing formation is exposed to a UV lamp, the fluorescent properties can identify the type of oil present. Different colors of fluorescence indicate different grades of oil. Moving from low to high API Gravity[1], colors seen will be brown (below 15° API), orange, yellow, white, blue-white, and violet (45° API).

On modern drilling operations, specialist contractors are used to monitor and analyze surface data and samples. This activity is called "mud logging".

Mud logging

Mud Logging involves taking measurements and samples during drilling. This usually includes

1. Taking samples of cuttings, mud, and formation fluid shows[2] and analyzing them
2. Logging and recording all important drilling parameters
3. Detecting and warning of the presence of problems such as kicks, H_2S, and washouts
4. Producing analyses and reports
5. Recording quantities and descriptions of cavings

A fully computerized unit with readout screens in the Drilling Supervisors office is the "Rolls Royce" of mud logging systems. With experienced engineers providing round-the-clock cover while drilling; warnings can be provided to the drillers of impending problems (such as pore pressure increases, wellbore instability, etc.).

The unit computers record a range of parameters relative to time and depth. At the end of the well these are handed over electronically to the Operator. Typical recordings will be made of

- Depth in feet or meter intervals
- Drilling rate, both in minutes per foot (meter) and in feet (meters) per hour
- Weight on bit
- Rotary speed
- Rotary torque
- Pump output
- Pump pressure
- Mud density being pumped into the well
- Mud density returning from the annulus
- Mud temperature in
- Mud temperature out
- Levels of gas dissolved or present in the mud returning from the annulus

If problems are encountered (such as stuck pipe or a break in the drillstring), this recorded data is very useful in helping to determine the root causes of the problem. Only by knowing the root causes can a strategy be developed to solve the problem and avoid a future recurrence.

Physical Sampling Downhole

Coring is the act of retrieving a whole sample of the downhole formations for analysis at the surface. Several classifications of coring can be made

- Bottom hole coring
- Explosive sidewall coring
- Rotary sidewall coring

Each of these is described in more detail.

Bottomhole coring

A special drill bit with a hole in the center cuts a doughnut shaped hole. The column of rock sticking up inside the core bit is protected by an arrangement of tubes. When coring is complete, the coring assembly is pulled out to the surface and the core is held within. This is the most expensive type of coring, but gives the most useful sample for analysis.

There are many different bottomhole coring systems available. Choosing between them will depend on the formation being cored (whether complete and well consolidated, unconsolidated, or fractured); what the hole inclination will be and what state the core should be recovered in (it can be kept under downhole pressure, if required).

Often on an exploration well, core equipment is kept ready on the rig, with an instruction to "core on *shows*" (a show is crude oil in the mud), as well as to core once the potential reservoir is reached.

Recovering a whole piece of formation allows some properties to be measured that cannot be adequately measured by logs. Permeability tests can be done using plugs of formation by flowing fluids under pressure through the core plug. Magnetic Resonance Imaging (MRI) can be used on the core to show the internal structure of the core. Many other useful tests can be carried out on cores, which can justify the cost and difficulty of coring.

Some of the routinely available coring systems include

- **Sleeve coring**. An outer steel barrel supports the core system. Inside this is an inner sleeve, which may be made of fiberglass, aluminum, or rubber. The inner sleeve supports loose, fractured, or unconsolidated formations. Once at surface, the core may be kept in the inner sleeve for transporting to the laboratory for analysis. Fiberglass or aluminum sleeves are often cut into convenient lengths (with the core inside), end caps installed to seal the ends, and boxed up for transport to the laboratory for analysis.

- **Sponge coring**. An aluminum inner sleeve has a sponge sleeve inside the aluminum tube. The core sits inside the sponge sleeve. When the core is recovered to the surface, any formation fluids that bleed out of the core are absorbed by the sponge where they can be later analyzed.

- **Orientated coring**. A knife blade creates a scratch mark along the core, showing downhole orientation. This can be very important if permeability is highly directional, or if the bedding

- **Pressure coring.** After cutting the core, the core barrel is sealed downhole so that when the core reaches the surface, it is still kept at the pressure of it's downhole environment. Of course, as the core barrel cools down during recovery, the internal pressures will reduce somewhat, which cannot be avoided. However, if the core contains dissolved gases or very light hydrocarbons, these are kept in solution during recovery. Pressure coring is very expensive. The system is sent to the wellsite inside a standard 40' oceangoing container.

Successful coring requires a lot of planning and coordination. The drilling fluid may be required to have special physical or chemical attributes in order to help avoid contamination of the core and to preserve it as it is pulled out of the well. Surface handling, preservation, and storage are extremely important to ensure that the core is in good condition by the time it arrives at the core analysis laboratory, as well as to recognize and document important information with the core on the rig.

Explosive sidewall coring

A tool is run in the hole on wireline that incorporates hardened, hollowed steel bullets inside a long steel carrier. The bullets are secured to the carrier with two wires. After placing the bullet at the correct depth for a sample, an explosive charge drives the bullet into the wall. The carrier is pulled up on it's cable, exerting a pull on the wires holding the bullet. With a bit of luck, the bullet will have penetrated to the correct depth, the wires will not break, and the bullet and sample will be recovered.

The bullet must be selected for the hardness of the formation to be sampled. The wrong choice may lead to the bullet bouncing off the wall (no sample recovered) or overpenetrating (the wire will break before the bullet can be pulled out of the wall).

With explosive sidewall coring, the sample will probably not give useful information on physical structures, such as fractures or bedding planes. The force of the bullet hitting the wall will fragment the formation. Information usefully gained will include porosity and permeability, confirmation of hydrocarbon presence, determination of clay content, and grain density and lithology.

Chapter 10 • Evaluation

The explosive sidewall-coring tool may carry up to 90 sampling bullets. Different bullets can be installed to match the anticipated hardness of the formations to each bullet.

Fig. 10-1 Explosive Sidewall-coring Bullet.
Courtesy of Schlumberger

Rotary sidewall coring

A tool that incorporates a small rotary core barrel and diamond bit is run in the hole on wireline. When deployed, the core barrel moves out of the side of the tool to contact the formation and starts to turn. It drills into the formation and recovers a standard sized plug of formation, 2.4 cm in diameter and up to 4.4 cm in length. Up to 50 samples can be taken in one trip in the hole, with the samples being stored inside the tool. The major advantage over explosive sidewall coring is that a relatively undisturbed core sample is recovered. Internal structures and fractures can be seen.

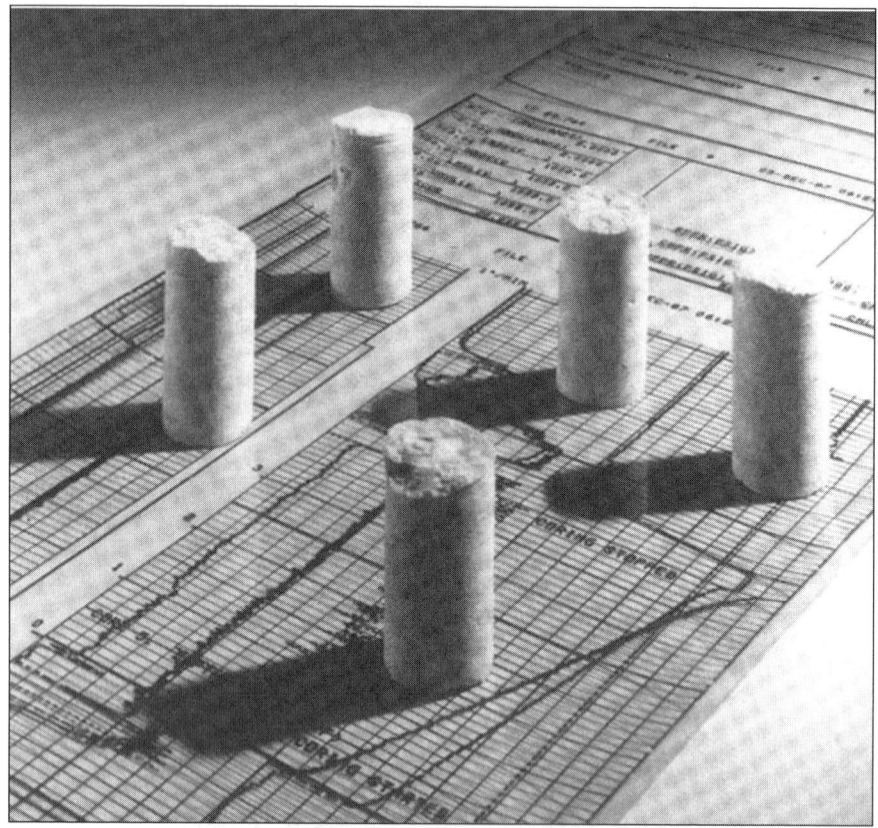

Fig. 10-2 Cores taken with a rotary sidewall-coring tool. Courtesy of Schlumberger

Pore fluid sampling and pressure testing

A tool may be run on wireline to a permeable formation of interest. Once on depth, the tool is 'set' by extending a probe until it contacts the borehole wall. A seal around the probe isolates the tool from the surrounding drilling fluid. Once the seal is established it is possible to measure the pressure of the pore fluids. Depending on the actual tool and configuration, samples of pore fluid may be obtained from one or more formations and recovered to the surface. The data recorded during sample taking may be used to calculate permeability at the point where the probe contacts the formation.

Fig. 10-3 Modular Dynamics Testing tool for sampling fluids & pressures. Courtesy of Schlumberger

Electrical Logging

The first electrical log was run in France by Conrad and Marcel Schlumberger in 1927, after they had experimented with measuring electrical properties of the Earth's crust. Their electric log measured the electrical potential between the surface of the Earth and a probe lowered into a well. This "Spontaneous Potential" log allowed the identification of different downhole formations by the depth that the electrical potential was seen to change.

Since that first log, Schlumberger and their competitors have developed many different techniques for measuring downhole physical, chemical, radioactive, and electrical properties. Taken together, these measurements can identify various lithologies, pore fluids, hydrocarbon presence, and other characteristics.

Electric logs are normally run on a special wireline that contains electrical conductors within it. In difficult wellbore conditions (high inclination, rough wellbore, and potential sticking conditions) the logging tools can be attached to the bottom of drillpipe or coiled tubing with the electrical cables run inside the pipe. Logging tools are also available that either transmit the information to the surface using pressure pulses in the mud while drilling, or that can record information within the tool for downloading to a computer at surface. These will be described in more detail later in this chapter. However, the operating principles of the various tools still apply, regardless of how the tools are run in the hole.

Below are described the major classes of electrical logging tools and the basic operating principles of each.

Resistivity and Induction tools. The electrical resistivity of a formation is related to the amount of water contained within the formation (due to the porosity) and the electrical resistivity of the formation water. Most sedimentary rocks do not conduct electricity when water is not present within the rock.

Resistivity logs. Measure resistivity directly by passing a current between electrodes touching the formation. This requires a conductive mud (i.e., water based, not oil- or gas-based) to work.

Induction logs. Measure formation resistivity indirectly by inducing an electrical flow in the formation using one coil and measuring the induced current with another coil. Induction logs work well in oil-based mud, where Resistivity logs do not.

Microresistivity tools. Measure electrical resistance with a very fine resolution at several different places around the circumference of the wellbore. These can produce a colour-coded picture of the wellbore with the color varying by resistivity. It is possible to measure the "dip" of a formation with a microresistivity tool, and as these tools also incorporate a north sensing tool, the direction that the formation dips can also be measured. Measuring resistivity helps to locate porous formations, show boundaries between formations, and identify hydrocarbon bearing zones.

Sonic tools. A pulse of sound is transmitted at the lower end of the tool and the time it takes the sound to travel a known distance through the rock is measured. Sonic tools can measure formation densities, compressive strengths, and can identify formation fractures. Information on permeability can be obtained. When run inside the casing, sonic tools can determine the quality of the set cement outside of the casing and can measure the inside diameter and thickness of the casing–identifying areas of casing damage. Sonic tools give much information of direct interest to the drillers. One such log is shown below–this is a tool used to evaluate the quality of cement after it has set outside of the casing. The log is depth-based on the vertical axis. In the left track on the log are some raw measurements and depth correlation information, such as a Gamma Ray log. (GR can be read through steel casing and this can be used to check that the depths are correct by comparing this log with earlier logs including a GR track). In the next track (2nd from left) are the sonic velocities and other sonic data. In the 3rd track

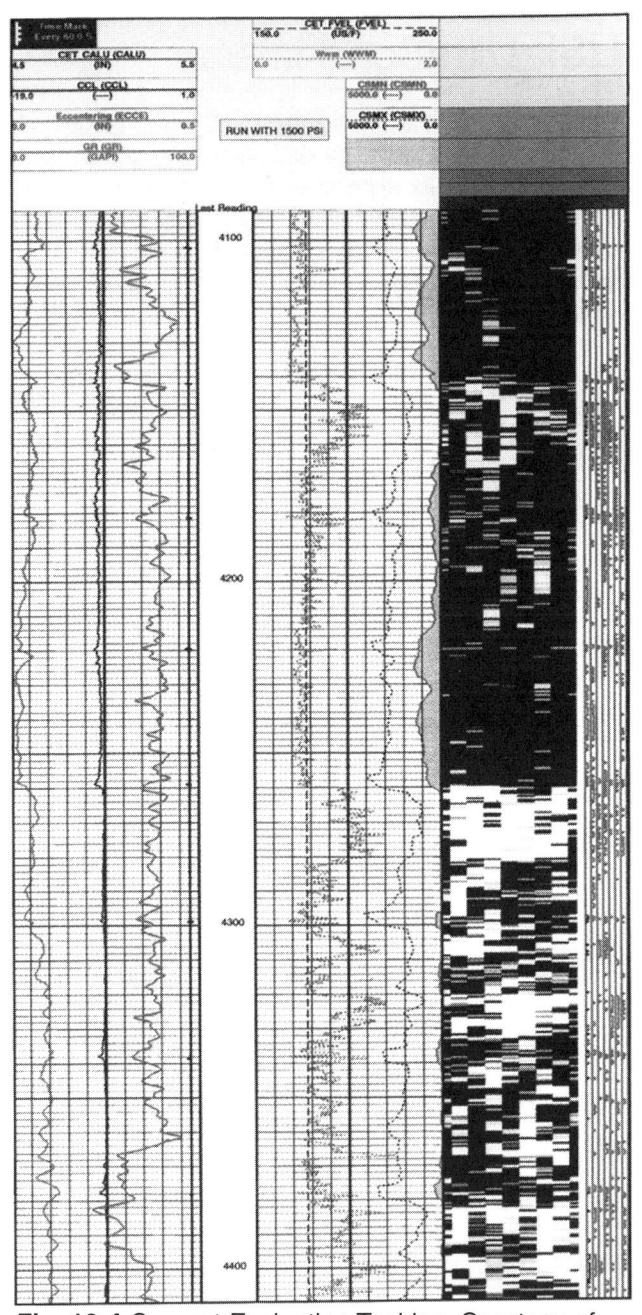

Fig. 10-4 Cement Evaluation Tool log. Courtesy of Schlumberger

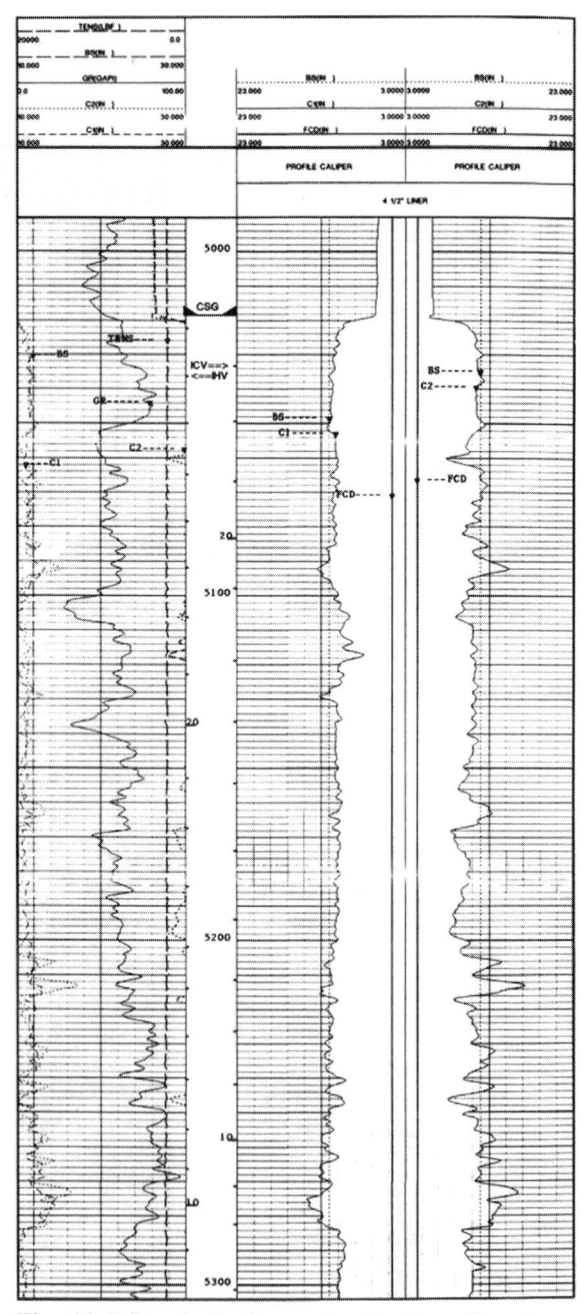

Drilling Technology in Nontechnical Language

Fig. 10-5 Borehole Geometry Tool log. Courtesy of Schlumberger

is a color-coded log directly indicating how good the cement is. This makes interpretation by non-experts relatively simple.

Radioactivity tools. These measure the natural radioactivity of the rocks (gamma rays). Some tools can even tell what kind of mineral is responsible for the radioactivity; identifying different shale types, for instance. Other radioactive tools bombard the formations with neutrons and measure the response from the formation, which can identify the amount of Hydrogen in the formation fluids (that would, for instance, identify the presence of hydrocarbons). Using tools that contain radioactive sources requires stringent safety precautions to ensure that personnel are not subjected to damaging levels of radiation. Most authorities around the world also mandate that any radioactive sources lost down the hole cannot be abandoned but must be recovered, even if the expense of recovery is very high. Special permission is then needed to abandon it if it cannot be recovered even with great effort.

Mechanical tools. These deploy arms that press against the wellbore and measure the diameter as the tool ascends. These "Caliper" tools may deploy one arm, two arms, or four arms for measuring the open hole. A four-arm caliper is especially useful because it gives a good indication of the wellbore profile. A wellbore may be the same size as the drillbit (called "in gauge"), it may be uniformly enlarged ("over gauge"), or it may be enlarged more in one direction than another. The type and extent of hole enlargement is an indication of how stable the wellbore is, as is the size and direction of the highest stress in the rock. A caliper tool also allows the annular volume to be accurately calculated, so that the correct volume of cement can be pumped around the casing.

If the drillstring becomes stuck, it is important to know at what depth the string is stuck. A tool called a "Free Point Indicator" can be run inside the pipe that anchors itself to the inside of the drillpipe at two places. The driller then applies pull to the pipe and the small amount of stretch in free pipe can be detected by the FPI tool. The driller then applies torque at the surface and the small amount of twisting in the free pipe can be detected. By moving the tool to different places in the drillpipe, it is possible to detect where the pipe is free in tension and in torsion.

Once the pipe depth is known, above which the pipe is free, it is then possible to apply left-hand torque in the string (to unscrew the pipe). A small explosive charge is set off on wireline as low as possible within the free pipe; the shock and vibration of this explosion will allow one (or more) of the drillpipe connections to unscrew. This process is called an "explosive backoff".

Electrical Potential tools. Also called "Spontaneous Potential", they measure the voltage arising between a formation downhole and the surface. This was the first kind of log run, as mentioned above.

Temperature log. Measures the temperature of the well, which varies with depth. Temperature has a large effect on the setting time of cement, and it must be accurately known when designing and testing cement slurries.

The Geologists and Reservoir Engineers will specify what logs they want to run in each hole section, so that they can further identify interesting formation characteristics. The reservoir will have many logs run through it, so that as much as possible can be known about the reservoir and the fluids it contains. In addition, the Drilling Engineers should add their own logging requirements to the program to ensure that they get the information they need to drill the next wells more cost effectively.

Conventional wireline logging

A standard logging cable is a wound wire rope with a diameter of 9/16". Instead of a rope core in the center of the wire rope, it contains a set of electrical conductors that transmit power to the logging tools and data back to the surface.

On the rig, a *logging unit* consists of a cabin that may be mounted on skids to be sent to offshore rigs by boat, or it may be installed on a truck. On a standard unit, the wireline winch contains 25,000' to 30,000' of cable. Within the logging unit are controls for the winch (to lower tools in and bring them out of the well) and a powerful computer network that analyses the signals from the tool while displaying the results on screen and printing them out to continuous paper.

The logging unit may have self-contained satellite communications, or it may use the rig communication system to transmit log results back to the Operating Company. Sometimes, important decisions must be made as soon as possible after logging, and the rig may have to wait for those decisions to be made before operations continue. With total daily operating costs that can exceed six figures in some cases, clearly the time taken to make these decisions must be minimized.

Chapter 10 • Evaluation

On an exploration well, where the hole is drilled solely to obtain information, the logging program for each hole section (and especially in the reservoir) will be extensive. Money invested in logs at this stage allows better decisions to be made when designing and drilling subsequent wells in the field. The logs will also allow any zones bearing hydrocarbons to be identified and to quantify some basic reservoir properties such as porosity and permeability. The major objective is to identify whether produceable hydrocarbons are present in commercially viable quantities. If so, then further wells will normally be drilled to extend knowledge of the reservoir before wells are drilled to produce the hydrocarbons. These further wells are called "Appraisal" wells.

If the hole condition is good (stable, in-gauge wellbore with no particular hole problems) it is possible to run a log on wireline at inclinations up to 50°. If the hole is badly enlarged in places, has potential sticking problems, or has a higher inclination, then other methods of deploying the logging tools have to be considered. These can add greatly to the cost of obtaining logs.

Logging using drillpipe or coiled tubing

If the wellbore passes through some problem sections, but the reservoir is loggable with standard techniques, drillpipe that is open at the lower end can be run to the point where the hole is in good shape. Special, slim (small diameter) logging tools can be run on wireline through the drillpipe, exiting at the bottom of the pipe, and then logging conventionally. In this case, the drillpipe merely acts as a conduit to get the logging tools past the problem areas. Many wells build up to a high inclination but then drop angle back towards vertical through the reservoir; if the maximum inclination is too

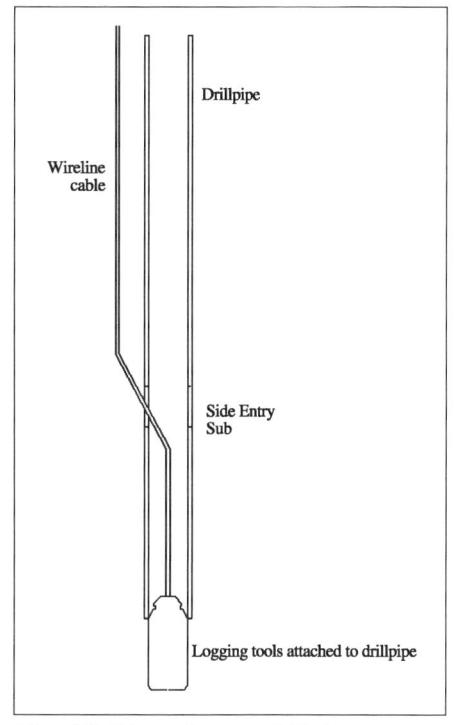

Fig. 10-6 Logging on drillpipe

231

high for the tools to slide down the drillpipe, then mud can be pumped in at the top to push the tools along the pipe.

For more severe conditions where the wellbore through the formations to be logged cannot be logged conventionally, the tools can be physically attached to the bottom of the drillpipe. The pipe, with the logging tools below, is run in the hole for a distance. Wireline is run inside the pipe with a special connector on the bottom, which latches onto the top of the logging tools. A special sub, called a *Side Entry Sub*, is then used to allow the cable to be run *outside* the pipe to surface. Now, as the drillpipe and tools are run in and pulled out, the winch operator can pay out or pull in the wire to keep the wire in tension.

Logging while drilling (LWD)

In the last ten years or so, logging sensors have been built in to drillcollars that are robust enough to withstand the heat, shock loadings, and vibration of drilling. These tools record log data while drilling progresses.

Measurement While Drilling (MWD) tools were described in Chapter 8. If log data is required in real time at the surface, the telemetry system of an MWD tool is used to transmit the data to the surface using mud pressure pulses. This requires that both an LWD tool and an MWD tool is included in the BHA and a physical connection using electrical wiring is made between the two tools. In a deviated well it may be necessary to transmit MWD data, too, so that the bit can be accurately navigated to the target.

LWD tools can also be run in "record" mode. Data is recorded within a memory module inside the tool, and when the tool returns to the surface this data is downloaded to a computer for analysis and printing.

LWD tools have developed to the point where the data they provide is accurate and reliable enough to replace wireline logging tools. The cost of using LWD is then partially offset by the time and cost saving of avoiding wireline logging. LWD can also obtain data in conditions where wireline tools cannot be run. If navigating within the reservoir (say in a horizontal well) then the LWD data becomes important to ensure that the wellbore stays in the correct place.

Production Testing

While evaluation of cuttings, mud, drilling parameters, and electrical logs is vital to understand the static reservoir characteristics, the only way

Chapter 10 • Evaluation

to ascertain *dynamic* reservoir performance is to let the well flow. An exploration well that locates possibly commercial quantities of hydrocarbons is usually tested by flowing it for periods of time and measuring the response (pressure) of the reservoir. This allows the operator to build a model to predict production rates and total hydrocarbon volumes for the reservoir under different operating conditions.

Modern reservoir testing has evolved from simply "seeing how much the well could produce" into a sophisticated tool for evaluating the reservoir. Information that the test can produce include

1. Measuring of the rate of flow and the corresponding pressures
2. Sampling the reservoir fluids and gases
3. Measuring the reservoir static temperature and pressure
4. Evaluate the extent of damage done to the reservoir by drilling into it
5. Identifying internal reservoir boundaries
6. Measuring the total deliverability of the well (maximum flow rate)
7. Identifying a flow rate at which sand particles start to move into the wellbore
8. Quantifying the effect of stimulation[3] work done to the well

A well production test records the downhole pressure response over time to changes in the flow rate. Very accurate pressure gauges are placed in the bottom of the well and the well is allowed to flow through an orifice of fixed size. This orifice is called a "choke". It will take some time for the flow to stabilize at a steady rate. Once the flow rate is stable, the well will be flowed for a period of time. After that period of time, the well will be shut in and the pressure inside the well will build up, again, over a period of time. These pressures are measured and plotted on a graph of pressure vs. time, as shown below.

The time taken for the well to reach a stable drawdown indicates the permeability of the reservoir. The greater the permeability, the faster a stable drawdown is reached. However, analysis of the drawdown data is difficult because the flow rates are not stable until drawdown stabilizes. It is therefore preferred to analyze the buildup curve once the well is shut in

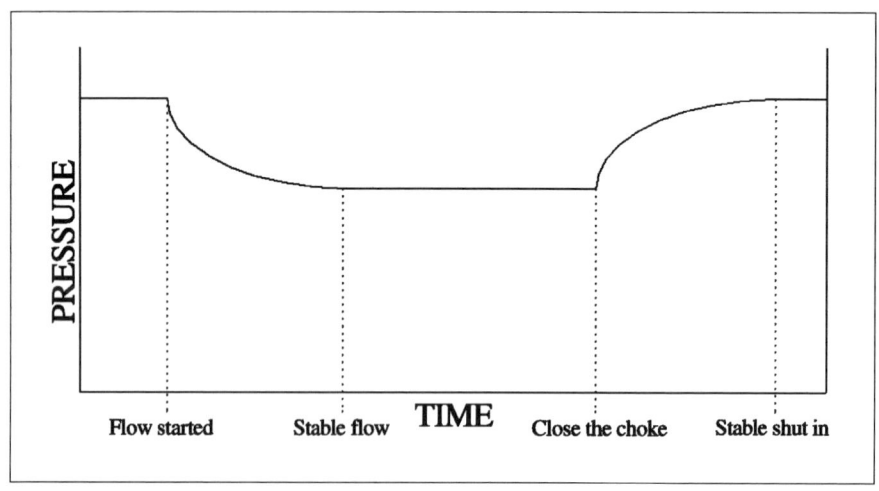

Drilling Technology in Nontechnical Language

Fig. 10-7 Graph of wellbore pressure vs. time during drawdown

again to evaluate permeability and wellbore damage. The faster the pressure in the well builds up to reservoir pressure, the higher the permeability.

If the well is flowing at a stable rate and a sudden change in rate is made (by changing the choke size); a pressure disturbance is created within the reservoir. This disturbance moves like a shock wave away from the wellbore through the reservoir. This shock wave may be reflected off internal disruptions to the reservoir, or off the outer boundaries of the reservoir. If it hits a gas cap, the shock wave may simply dissipate. Any reflections can be recognized by a change in the measured pressure once the reflected wave comes back to the wellbore. This type of test is called a "Transient test". Transient testing, made possible by the extreme accuracy of modern gauges and very powerful computer systems, allows well test interpretations that provide a description of the internal geometry of the reservoir.

To analyze a transient test, two curves are produced of pressure vs. time using log-log axes. One curve simply plots pressure against time ("pressure curve"), the other plots the rate of pressure change against time ("derivative curve"). The shape of the derivative curve identifies features that would be too subtle to be recognized from the pressure curve alone. Early transient curve shapes were compared to a library of curves which were characteristic of various types of reservoir, but, by using computers, it is possible to compare a vast number of reservoir model shapes to the observed data.

Chapter 10 • Evaluation

The plot of pressure and derivative curves makes it possible to identify many more reservoir characteristics than was previously possible. Some examples are shown below–the lower curve in each graph is the derivative, the top curve is the pressure.

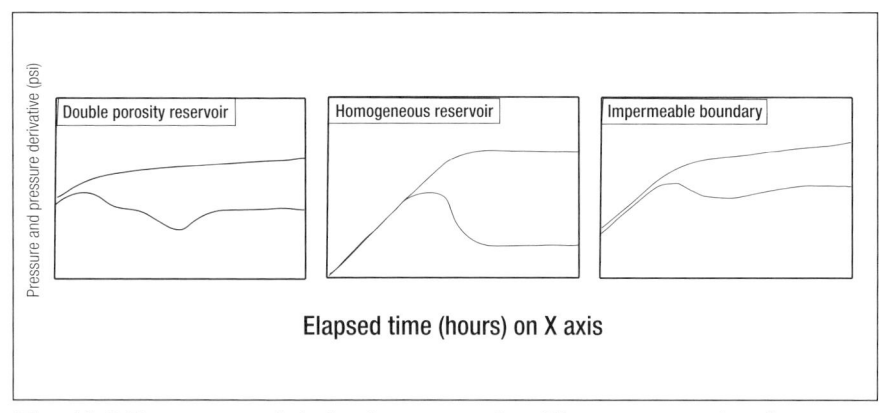

Fig. 10-8 Pressure and derivative curves for different reservoirs. Courtesy of Schlumberger

Chapter Summary

This chapter showed how data from the well is obtained using various methods of sampling and measuring. This data is used to monitor and optimize performance; optimize future drilling activities (well planning and operations), recognize a commercially viable hydrocarbon reservoir, and to optimize exploitation of the reservoir. In particular, the activities encompassed by mud logging, wireline logging, surface and subsurface sampling, and production well testing were discussed in some detail. The working principles behind different types of electric logging were described as well..

Glossary

[1] **API Gravity.** The density of an oil is expressed in Degrees. An API gravity of 10° is equal to freshwater density (Specific Gravity of 1). An API gravity of 45° is equal to a Specific Gravity of 0.801. The higher the API gravity, the lower the SG and the more valuable the oil. This is because lighter oil contains more light hydrocarbon elements that require less refining to produce petroleum.

2 **A show.** An indication of hydrocarbons downhole, usually seen by examining drilled cuttings for fluorescence.

3 **Stimulation.** Mentioned briefly in Chapter 4. Stimulation refers to work performed on a well in order to increase the production potential (how fast hydrocarbons will flow into the wellbore at a given pressure drop). Techniques include deliberately fracturing the formation using high pressures and pumping acid to create larger channels.

Chapter 11

Well Control

Chapter Overview

The principles of hydrostatic pressure were described in detail at the end of chapter 1. Also mentioned was the result of drilling into a formation where the pore pressure exceeded mud hydrostatic pressure—a kick may occur where formation pore fluids enter the wellbore.

This chapter will define well control (primary, secondary, and tertiary). The processes and equipment involved in kick detection and control will be described. Special well control situations (shallow gas, kicks and losses, blowouts, high pressure high temperature wells, and underbalanced drilling) will also be discussed in sufficient detail to impart a basic understanding of the causes, effects, and implications.

Primary, Secondary, and Tertiary Well Control

The term *well control* refers to the control of downhole formation pressures penetrated by the well. There are three distinct well control levels that may occur during drilling operations.

Primary well control. The first line of defense. Primary control is given by maintaining the density of the drilling fluid such that hydrostatic pressure at all depths where formations are exposed exceeds formation pore pressures.

Mud hydrostatic pressure > Formation pore pressure

The well is planned and drilling operations are controlled with the intention that primary well control is *always* maintained. The exception to this is in underbalanced drilling, which is discussed later in this chapter.

When a kick is taken, primary control has been lost for some reason. There are four main reasons, in general, why primary control might be lost during drilling operations:

1. The well penetrates an overpressured zone with a higher formation pressure than mud hydrostatic.
2. A weak downhole formation allows sufficient mud to leave the wellbore so that the level of mud in the annulus drops. As hydrostatic pressure = gradient x depth (of the fluid column), if the top of the column drops then hydrostatic pressure along the wellbore decreases. If it drops far enough, hydrostatic overbalance on a permeable formation may be lost.
3. Failure to fill the hole properly when pulling out of the hole. As steel is pulled out of the well it has to be replaced by mud. If the driller does not keep the hole full while pulling out, the mud level in the annulus will drop and hydrostatic pressure will reduce.
4. Swabbing. Swab pressures were described in chapter 7. If the drillstring is pulled up with sufficient speed, the reduction in pressure at the bottom can be enough to allow formation fluids to enter the well.

If primary control is lost and formation fluids start to flow in to the well, *secondary well control* is initiated by closing the BOP to seal off the annulus. As fluid enters the well from the kicking formation, pressure in the well will increase until the total pressure exerted by the mud on the kicking formation equals the formation pore pressure. The pressure exerted by the mud equals mud hydrostatic pressure + surface pressure held by the BOP.

Mud hydrostatic pressure + surface pressure = Formation pore pressure

As the mud hydrostatic pressure and surface pressure are both known, the formation pore pressure can be calculated.

The objective now is to restore primary control. This is achieved by circulating mud of higher density into the well so that mud hydrostatic pressure again exceeds formation pore pressure without any pressure held on surface by the BOP.

It sometimes happens that the BOP equipment fails, or the hole starts to allow fluid to leak into an underground formation. Secondary control cannot be maintained and formation fluid again starts to enter the wellbore. This is now a dangerous situation calling for extreme measures to restore control. If control is not restored, the end result is a blowout. *Tertiary control* has to be applied to try to stop the flow.

Tertiary control involves pumping substances into the wellbore to try to physically stop the flow downhole. This may involve pumping cement (with a high risk of having to abandon the well afterwards). However, there is another method that may be employed, called a barite plug.

A barite plug is set by mixing a heavy slurry of barite in water or diesel oil. It has to be kept moving while mixing and pumping. Once the slurry is in position downhole and pumping stops, the barite rapidly settles out to form an impermeable mass that will hopefully stop the flow of formation fluid. The main risk is that if pumping stops with the slurry inside the pipe, barite will settle out in the pipe and plug the drillstring.

BOP Stack

When planning and drilling wells, the assumption is made that a *kick is always possible*. Even if the well is the 100th drilled in the immediate area, primary control can still be lost for some reason. This is why BOPs are always used once the surface casing has been cemented in place.

The primary function of the BOP is to form a rapid and reliable seal around the drillstring or across the empty hole (if no pipe is in the hole) to contain downhole pressures. There are currently two types of preventers available that allow this seal to be formed. Most BOPs contain at least one of each type as the different characteristics of each are useful for different operations.

Bag type preventer (called *annular preventer*)

A bag preventer contains a large circular (viewed from above) rubber element, conical if viewed from the side. This is held inside a steel chamber. Below the rubber element is a hydraulically operated piston. As the pis-

Drilling Technology in Nontechnical Language

ton moves up, the rubber element is compressed and pushed inwards. The element can distort to allow it to seal around any smooth object in the wellbore, whether it is a round pipe or a square kelly. It can also seal on the open hole, but the level of rubber element distortion necessary will seriously shorten the useful life of the element Fig. 11-1).

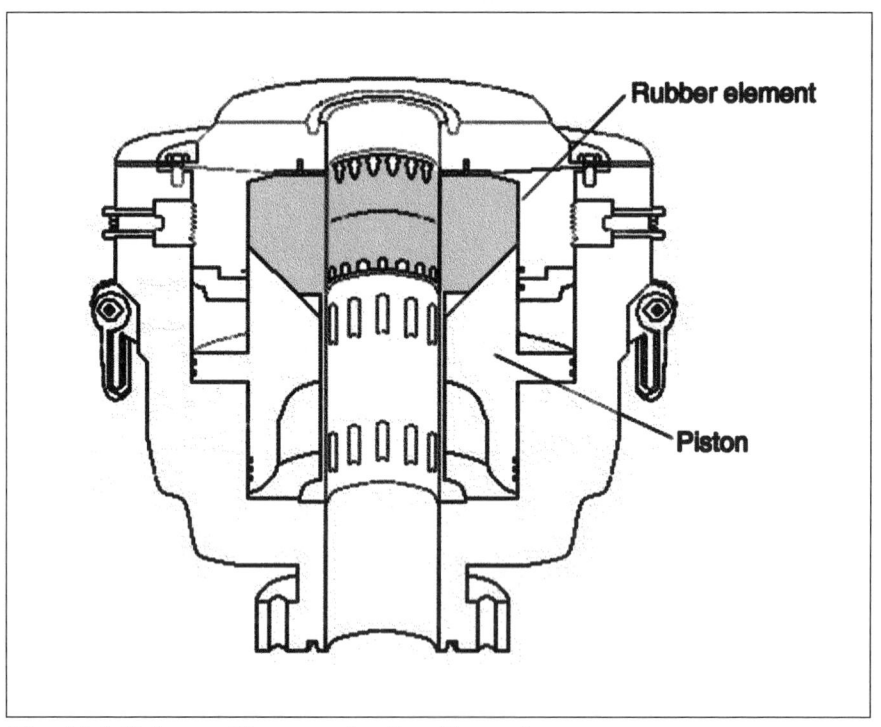

Fig. 11-1 Cutaway Diagram of a Bag Preventer

Apart from sealing on irregular shaped tubulars, the annular preventer can also be used to allow pipe movement in or out of the well under pressure. Moving pipe in a closed well when the well is under pressure is called *stripping*. This may be useful if a kick occurs when little or no pipe is in the hole. In order to kill the well, heavy fluid has to be pumped through the drillstring to the bottom of the hole. If the pipe is not deep enough, more pipe can be added until it is deep enough to kill the well by *stripping* pipe through the bag preventer. This is possible because the rubber element can move to accommodate the thicker tool joints moving down through the element (Fig. 11-2).

Chapter 11 • Well Control

To strip in, the hydraulic closing pressure on the piston is reduced until a slight amount of mud starts to leak through the seal. This provides some lubrication. Grease is put onto the tool joints and the pipe is slowly moved downward to minimize wear on the rubber seal.

Fig. 11-2 Replacement Rubber Element for a Bag Type Preventer

Ram type preventer

The other type of preventer uses large steel rams that close together to press a rubber sealing element against the pipe. These rams are interchangeable and are of the following three kinds:

1. **Fixed pipe ram.** The seal is sized to fit one outside diameter only (Fig. 11-3).

2. **Variable bore pipe ram.** The seal element can accommodate a narrow range of diameters, for example 3-1/2" to 7". It can only seal on round pipe, not square or hexagonal shapes (Fig. 11-4).

Drilling Technology in Nontechnical Language

Fig. 11-3 Set of Fixed Pipe Rams

Fig. 11-4 Variable Bore Ram (one of a set of two) Showing the Sealing Element

Chapter 11 • Well Control

3. **Blind rams.** Designed to seal on the open hole. Blind-shear rams have blades incorporated that can cut through drillpipe (though not through drill collars or casing) (Fig. 11-5).

Fig. 11-5 Blind-shear Ram (one of a set of two) Showing the Flat Seal and the Blade Below

Normally, a BOP stack would have at least one bag preventer and two ram preventers, as shown in Figure 11-6. BOP stacks for deeper wells might have up to four ram preventers and two Annular preventers. The ram preventers generally have a higher pressure rating and are always installed below the bag preventers. If only two ram preventers are used, the bottom set will normally be blind-shear rams and the upper set will be pipe rams. One reason for placing the blind rams on the bottom is that if the pipe rams leak, it is possible to close the blind rams below and safely open up the upper preventer to change the rams.

Below the lower rams are pipes that come out to the side. These are called *side outlets* and are used to allow flow out of or into the annulus during well killing operations.

The side outlets have different names. One side connects to the standpipe manifold to allow flow to be directed into the annulus. This is called the *kill line*. The opposite side connects to a manifold of valves and chokes. This is called the *choke line* and its purpose is to control the flow out of the annulus. Chokes are described below.

Fig. 11-6 BOP Stack Assembly for a Land Rig (one bag, two ram preventers)

Once the driller detects that a kick is in progress, one of the BOP stack preventer units will be closed to seal the annulus of the well. The pressure at the top of the well (inside the drillpipe and at the top of the annulus) will be recorded. Once the pressures are steady, the formation pore pressure (due to hydrostatic + surface pressure) can be calculated. A plan can then be formulated to kill the well.

Choke valves

There is another vital item that forms part of the BOP equipment. A choke valve allows fluid to flow through it, but has a variable sized opening. This allows mud to flow out of the annulus but at the same time keep pressure on the annulus. (See Fig. 11-7)

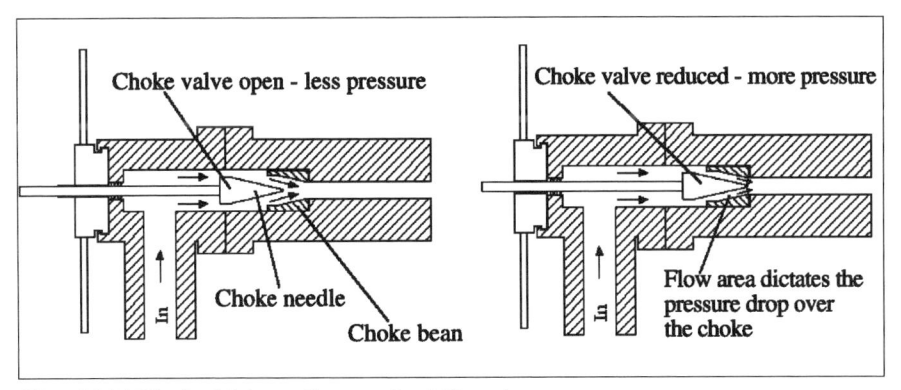

Fig. 11-7 Choke Valve—Conceptual Drawing

The pointed part of the needle moves in and out of the choke bean. As the needle moves in (to the right in the figure above) the gap closes. For a particular flow rate through the choke, this will increase the pressure upstream of the choke. How this is used during killing the well is described later in this chapter.

It would be possible to use a normal valve as a choke. However, mud that contains solids (barite, bentonite, sand, and other drilled particles) is quite abrasive when flowing through a restriction at high pressure. A normal valve would soon erode and fail to hold pressure if it were used to exert pressure on the flowing mud. A choke valve is designed to handle this operation with minimum erosion by its design and by the use of tungsten carbide internal components. It is still possible for the choke to become eroded during a well killing operation, so the rig must carry spares of these parts. There must also be valves positioned upstream of the choke that can be closed in order to allow the choke to be repaired.

BOP control systems

BOP units (bag and ram preventers) are moved using hydraulic fluid under pressure. To provide this pressure, a hydraulic control system is used that contains the following elements (Fig. 11-8):

1. A reserve hydraulic fluid tank holding fluid at atmospheric pressure.
2. A set of bottles holding fluid under high pressure (usually 3,000 psi) with pressurized nitrogen.

Drilling Technology in Nontechnical Language

3. A high pressure manifold connected to the bottle system.
4. A low pressure manifold that contains fluid at the working pressure of the ram preventers (usually 1,500 psi).
5. A pressure regulator that feeds fluid from the high pressure manifold to the low pressure manifold, which reduces the unregulated pressure to the working pressure.
6. A set of valves attached to the low pressure manifold that can direct working pressure fluid to the rams (to open or to close them) and that directs exhaust fluid back to the reserve hydraulic fluid tank.
7. A valve that controls the opening and closing of the bag preventer.
8. A pressure regulator that feeds fluid from the low pressure manifold to the bag preventer control valve. Bag preventers operate at a lower pressure than ram preventers (normally about 800 psi), but as the seal element gets worn, higher pressure is required to make the bag seal around the pipe. This regulator is also used to lower the closing pressure to the bag preventer if it is desired to strip pipe in through the preventer, as described previously in this chapter, or to increase the closing pressure on a worn element.
9. Two sets of pumps to maintain system pressure. One set is driven by compressed air from the rig air system, the other is powered by electricity. This provides some backup in the event of either the rig's air or electrical system failing. The air driven pumps give a higher flow rate and pressure the system to around 2,800 psi. The electric pump tops the system to the full pressure of 3,000 psi at a lower flow rate.

Subsea BOP systems

On floating rigs, the BOP is attached to the top of the surface casing at the seabed. In chapter 4, the process of drilling on a seabed template was discussed along with the concept of subsea wellheads and BOPs.

A subsea BOP contains extra control systems when compared to a surface BOP. The hydraulic control system is also *open ended*; hydraulic fluid is exhausted to the sea rather than being returned to the control system. The

Chapter 11 • Well Control

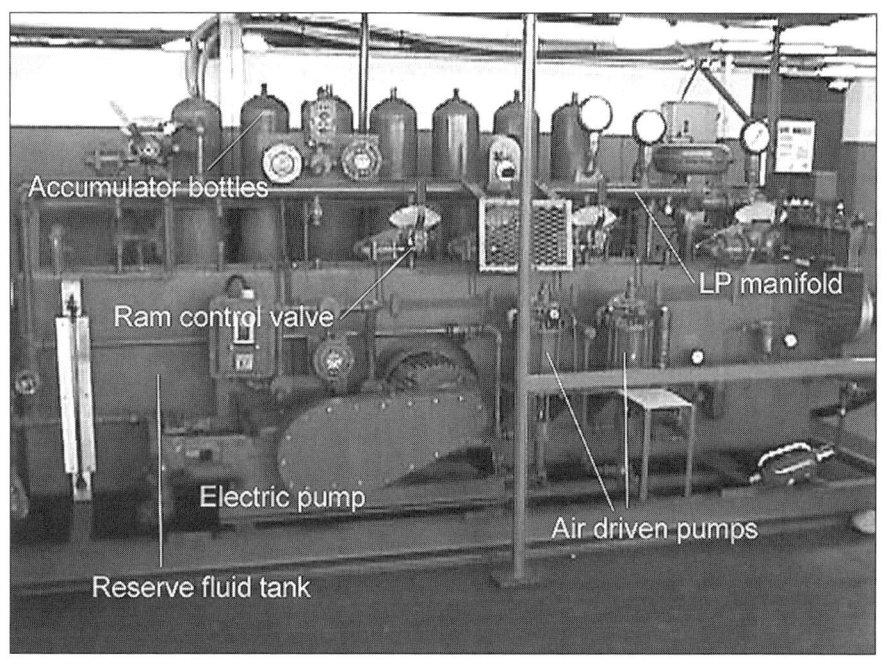

Fig. 11-8 BOP Control System for a Surface BOP Stack

hydraulic fluid in this case is water mixed with a non-toxic soluble oil so that pollution is avoided.

When drilling with a floating rig, it is important that the rig is able to move off location away from the well. This might be due to bad weather forcing operations to be suspended. The BOP system has to be capable of latching to and releasing from the wellhead, so at the bottom of the BOP is a hydraulic latch. Also the riser (described in chapter 4) must be capable of being released, leaving the BOP on the seabed and another hydraulic latch is placed here. As floating rigs (to some extent) do move, the riser and BOP system must accommodate some angular movement of the riser. This is achieved by incorporating a flexible joint below the riser. However, if angular movement exceeds a certain amount (usually around 5°), the rig will disconnect the riser from the BOP to prevent damage.

Kick Detection Equipment

When a kick occurs, there are almost always warning signs that come in advance of the actual kick. If the drillers (backed up by the mud loggers)

Drilling Technology in Nontechnical Language

are alert, they will be watching the well closely to take immediate action to close the well once a kick is recognized.

The following are two main kick detection systems that give a direct indication of a kick:

1. The pit-volume totalizer (described in chapter 5) provides a readout showing the total volume of drilling fluid held on the surface in the active system[1] tanks. If this total increases, and if the increase is not due to the mud engineer adding chemicals or fresh mud to the system, then a kick is occurring.

2. The flow indicator is an instrument attached to a paddle that sits in the flowline from the annulus. This paddle is pushed up by the returning mud stream—the amount it is pushed depends on the flow rate, among other things. If the flow rate should increase an alarm will sound. If the flow rate out increases, but the mud pump speed has not been increased, it is possible that the extra flow out is due to an influx entering the wellbore.

Generally, the flow indicator will give the first positive indication of a kick, followed by an increase in the active volume. However, the paddle type flow indicator is prone to false alarms because of cuttings and other debris that may stick to the paddle or build up underneath it.

If the surface instruments indicate that a kick is in progress while drilling, normally the driller will stop drilling, pick up the drillstring so that the bit is above the bottom of the hole and stop the pumps. A visual check is then made by looking down through the rotary table, into the bell nipple, at the level of mud in the annulus. If the well is in fact kicking, then the mud in the annulus will still be moving upwards even though the pumps are shut down. Having confirmed a kick, the driller will close the BOP as quickly as possible and notify the toolpusher and drilling supervisor[2] in charge of the rig.

If the permeability of the flowing formation is high, the kick can develop very quickly. A larger drilled hole will also allow influx to flow in faster but this is compensated for (to an extent) because the capacity of the hole is greater in a larger hole. It is also possible that if permeability is very low, little or no influx enters the wellbore even though mud hydrostatic is less than the formation pore pressure.

Chapter 11 • Well Control

Killing the Well

The operations involved in restoring primary control are known as *killing the well*. In principle it is simple, but in practice there are many considerations that are required to execute it safely. Problems may occur that have to be recognized and addressed quickly. The following example of a well kill will be worked through to describe the main points:

> A vertical well is being drilled at 8,000 ft. The bit size is 12-1/4", there are 300 ft of drill collars that have an OD of 8" and an ID of 3". Above the drill collars, 5" diameter drillpipe is in use with an ID of 4.276". Casing is set at 5,000 ft, this is 13-3/8" OD and 12.615" ID. The mud in use has a density gradient of 0.5 psi/ft.

A kick is taken and after closing the BOP, the driller has the following information about the kick:

1. The pressure inside the drillpipe at surface is 500 psi
2. The pressure on the annulus (inside the BOP) is 600 psi
3. The active mud system volume has increased by 21 barrels

In any problem situation, it's best to draw a simple diagram showing the main points. This helps to avoid mistakes. Figure 11-9 shows the diagram for this well.

The first thing to notice is that there is a difference between the pressure on the drillpipe and on the annulus. Although the pressure at the bottom of the well is the same, the fluid in the annulus now has a different hydrostatic pressure than the fluid in the drillpipe, because pore fluid (21 barrels of it) has entered the wellbore. This is lighter than the mud, so the hydrostatic pressure in the annulus is less (by 100 psi, the difference between drillpipe and annulus pressures).

Assuming that the hole is in gauge (the same diameter as the drill bit) and that the influx stays as a single column of fluid (doesn't mix with the mud), then the height of the influx in the annulus can be calculated. Twenty-one barrels in a 12-1/4" hole with 8" pipe inside it will have a height of 251 feet. Knowing the height of the influx, knowing how much hydro-

Drilling Technology in Nontechnical Language

static pressure was lost as a result of this (100 psi), and knowing the mud density, the density of the influx can be calculated.

$$\text{Influx density gradient} = \text{Mud gradient} - \frac{100 \text{ psi}}{251 \text{ feet}} = 0.102 \text{ psi/ft}$$

This density is consistent with an influx of gas. An oil influx would have a greater density (0.3 to 0.4) and a saltwater influx would be even more (around 0.47 psi/ft). Therefore the difference in annulus and drillpipe pressures, together with the volume of the influx, allows the probable type of influx to be calculated. In particular, it is important to identify whether the influx is gas or not because this effects the well kill operation. Gas is much harder to handle as it moves up the annulus because its pressure will drop. As the pressure drops, the volume will increase as predicted by Boyle's law[3].

Next, the pressure in the kicking formation can be calculated. Calculations of hydrostatic pressure were covered in chapter 1. As the fluid

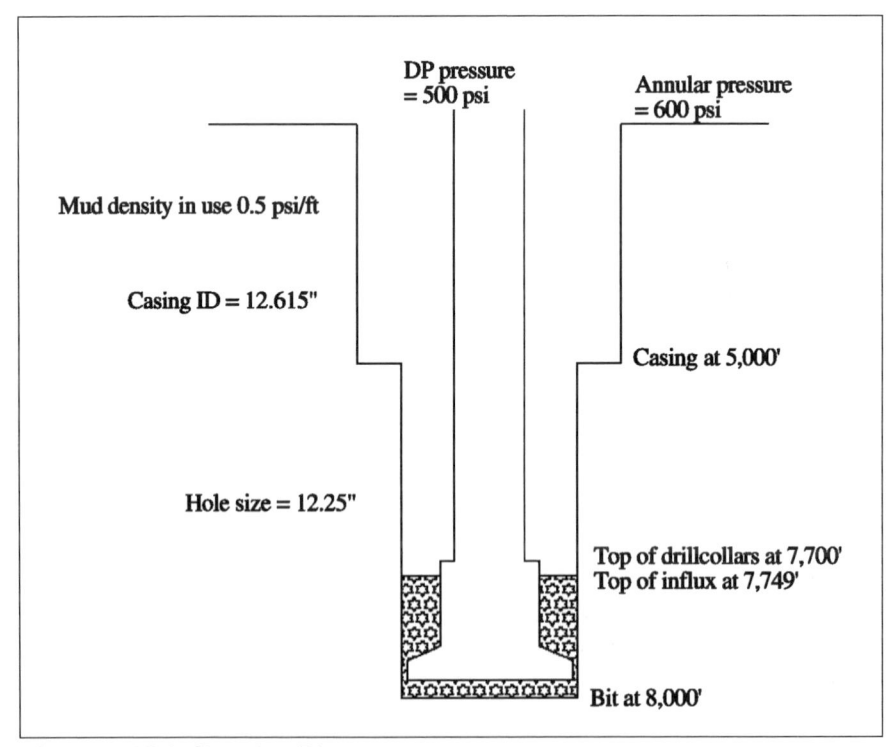

Fig. 11-9 Kick Situation Diagram

Chapter 11 • Well Control

inside the drillstring should not be contaminated with influx, the drillpipe pressure (Pdp) is used to calculate the bottom hole pressure, BHP (= formation pressure).

BHP = Hydrostatic Pressure + Pdp = (0.5 x 8,000) + 500 = 4,500 psi

To kill the well and restore primary control, heavier mud must be circulated into the well. To give a hydrostatic pressure of 4,500 psi at 8,000 ft, the kill mud density gradient (ρ_2)⁴ can be calculated.

$$\rho_2 = \frac{4,500}{8,000} = 0.563 \text{ psi/ft}$$

While preparations are made to kill the well, the mud engineer can start to add barite to the mud in the active system to increase the density gradient to 0.563 psi/ft. Meanwhile, some more calculations have to be made before the kill can start.

During a well kill operation, the intention is that bottom hole pressure should be kept as close to the formation pressure as possible. If the bottom hole pressure is allowed to drop below pore pressure then it is possible that more influx will enter the well, which will increase the complexity of the operation and increase the risk of problems. If the pressure rises very much, then a weak formation somewhere in the well might break and allow losses. If that happens then more influx will enter the well and will move into the loss zone. This is called an *internal blowout* and is much more complicated and dangerous than a simple killing operation, so it must be avoided.

The bottom hole pressure during the well kill is controlled by the choke on the annulus because applying pressure here causes pressure to be applied everywhere in the well, including the kicking formation.

The bottom hole pressure is monitored during the well kill by watching the pressure on the drillpipe (the pump pressure). The hydrostatic head of the mud in the annulus at any particular stage of the operation is known, so bottom hole pressure = pump pressure + hydrostatic pressure.

During drilling, the driller regularly makes a test of each pump by circulating at a set slow rate and measuring how much pressure is required to circulate around the well at that rate. All mud pumps are tested so any can be used for killing the well. Let's say that in this well, a slow circulating rate (SCR) of 30 strokes per minute was used. The pump outputs 5 gallons for every stroke, so at a SCR of 30, the pump output is 150 gallons a minute. There are 42 U.S. gallons in a barrel, so the flow rate is 3.57 barrels per minute. At this flow rate, the driller measured a pressure of 500 psi. This is known as the initial circulating pressure or PC_1.

When heavier mud is pumped into the system, more pressure is required to push this heavier mud around. If the mud is changed to heavier mud, the new pressure (denoted PC_2) can easily be calculated as

$$PC_2 = PC_1 \times \frac{\rho_2}{\rho_1} = 500 \times \frac{0.563}{0.5} = 563 \text{ psi}$$

This is the pressure that is then required to force the heavier mud to flow around the well. In fact the pressure required to give flow in the annulus is very small compared to the drillstring, so annular pressure loss is ignored and the assumption is made that PC_1 and PC_2 are the pressures required to give flow through the drillstring and drill bit only.

It can now be said that (ignoring the requirement to maintain bottom hole pressure equal to formation pressure) when circulating starts the pressure will be 500 psi and when kill mud reaches the bottom of the well, the circulating pressure will be 563 psi. Once heavy mud is at the bottom, hydrostatic pressure from the mud in the drillstring is sufficient to balance formation pressure. This phase of the kill operation, from starting to pump heavy mud until heavy mud reaches the bottom, is called *Phase 1*.

In order to maintain bottom hole pressure equal to formation pressure, extra pressure must be applied (by the choke) during Phase 1. Holding pressure on the choke will effect the pressure in the entire system, including the pump pressure. This extra pressure on the pump starts at the original drillpipe pressure (500 psi) and will decrease to zero by the time heavy mud reaches the bit. In a vertical well, this decrease is linear—for each barrel pumped, the extra pressure will reduce by the same amount. The volume inside the drillstring is known—it works out to be 139 barrels.

The initial circulating pressure equals the slow circulating rate pressure plus the shut in drillpipe pressure. Initial circulating pressure therefore is 500 + 500 = 1000 psi. Final circulating pressure is PC_2.

A table and graph of pump strokes against pump pressure for Phase 1 is in Figure 11-10:

Chapter 11 • Well Control

Pump Pressure	1000	956	913	869	825	781	738	694	650	606	563
Strokes pumped	0	117	234	351	468	585	702	819	936	1053	1170
Barrels pumped	0	14	28	42	56	70	84	98	111	125	139

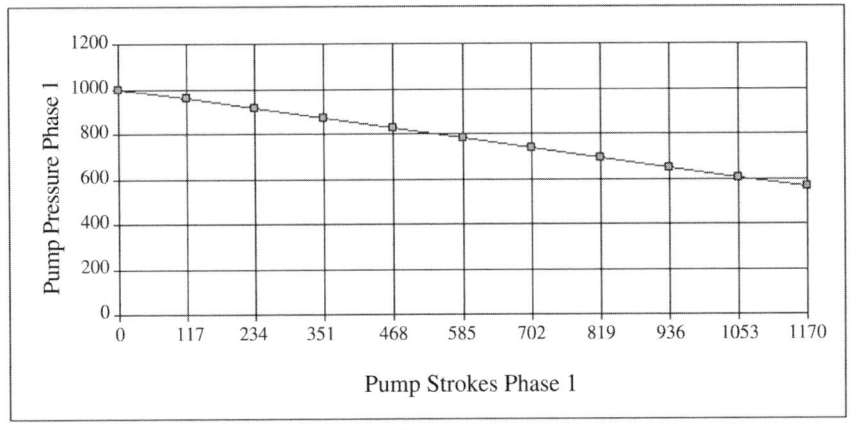

Fig. 11-10 Kill Phase 1 Pumping Pressure

Now once the surface active system mud is weighted up to ρ_2, the kill can begin. One person controls the choke valve and monitors both the drillpipe and the annulus gauges. As the pumps start and flow begins, the pressure on the annulus will start to rise. The choke is now opened to keep the choke pressure equal to the shut in choke pressure of 600 psi. This is maintained while the pump speed is slowly increased to 30 strokes a minute. Once the pump is at the correct speed, the choke operator needs to watch both choke and drillpipe gauges. Using the table or graph above, the choke setting is adjusted to keep the pump pressure correct. At this stage, pump pressure will naturally fall as heavy mud moves down the drillstring, and only slight choke adjustments are needed.

At the end of Phase 1, the pump pressure is then kept constant for the rest of the killing operation (in this case at 563 psi). The choke setting will change as the gas moves up the wellbore. As the gas moves up and expands in accordance with Boyle's law, it will push out more mud than enters the well. This reduces the hydrostatic pressure in the annulus and choke pressure has to be increased to compensate. Choke pressure will reach a maximum once the top of the gas reaches the choke and will decrease rapidly as mud replaces gas leaving the annulus.

Once the kill operation is complete, the pumps are stopped. If no pressure remains on drillpipe or annulus, the BOP is opened. A visual check is made to insure that the well is not flowing with the pumps off. If all is well, drilling can now be resumed—though in practice, in order to get the mud into proper shape for drilling, another full circulation is generally required.

There are other techniques for killing a well, depending on the circumstances. The technique outlined in this example is called *wait-and-weight* or the balanced method. Other methods not described here are drillers, volumetric, combined volumetric and stripping, and bullheading.

Shallow Gas

Shallow gas was discussed in chapter 3. The emphasis on shallow gas is prevention. If shallow gas is encountered and the well flows, the chances of stopping it are remote.

Instead of a BOP designed to close in the well, a diverter is used during drilling. This comprises a large bag preventer, underneath which are two large side outlets (up to 12" diameter) with some kind of full opening valve system. Once the well starts to flow, the side outlet facing downwind is opened and the bag preventer is closed. Flow from the well gets diverted downwind, away from the rig (Fig. 11-11).

Dynamic kill

Dynamic killing is a method for stopping an influx by increasing annular pressure losses (i.e., the pressure required to force mud to flow in the annulus). This pressure increase comes from pumping fluid as fast as possible into the well.

Dynamic killing of a shallow gas flow is unlikely to be successful except in rather narrow circumstances. To have any chance of working, a dynamic kill must have:

1. **Very high flow rates.** The maximum pump output on most rigs will be insufficient.

2. **Small hole diameter.** Once the blowout is established, large volumes of formation solids are blown out of the well so that the hole rapidly enlarges. This means that an attempt to dynamically kill the well must be made within seconds of a flow being identified.

Chapter 11 • Well Control

Fig. 11-11 Diverter and Side Outlets. Photo courtesy of Schlumberger

3. **Small annular clearance.** If large diameter drill collars and drillpipe are used, this reduces annular clearance and hence increases the annular pressure loss.

4. **Increased drilling fluid density and viscosity.** Surface holes are generally drilled with low density fluids because the formations are weak. Using high density, highly viscous fluid would increase the risk of mud losses into the formation. It is even possible to cause formations to fracture by using high density drilling fluid. This implies that a tank of heavy, viscous kill mud must be kept ready to pump at maximum rate as soon as the need is identified.

If a dynamic kill is the chosen method of killing a shallow gas blowout, it must be planned in advance to address each of these elements if there is to be any realistic prospect of success. Crew training is vital as is implementing equipment and procedures to detect an influx as early as possible.

Ultimately if the well is not killed quickly and the blowout develops, the chance of equipment failure is high. The extremely erosive nature of the flow (gas plus formation solids being an effective sandblaster) means that sooner or later holes will appear in the diverter system, allowing the flow to enter the rig. H_2S is sometimes encountered in shallow gas flows, which presents an immediate and serious threat to life. It is to be hoped that the accumulation of shallow gas is small and after a short while, it depletes and the flow stops. It is also possible that chunks of rock coming up the wellbore get stuck and form a bridge, stopping the flow.

Shallow gas on floating rigs

The best place to be if a shallow gas flow occurs is watching from a safe distance upwind! The second best place to be is on a floating rig, drilling with returns to the seabed.

Surface holes on floating rigs are normally drilled without returns to the rig. After cementing conductor pipe in the seabed, the next BHA is run into the conductor using guidance from subsea video cameras. Mud and cutting returns from the annulus exit into the sea. If a shallow gas flow occurs and if this presents a danger to the rig (gas appearing on the sea surface close to the rig), the driller drops the drillstring and the rig moves away.

Sometimes on floating rigs, returns are taken back up to the rig. This might be the case in deeper water, or if government regulations prohibit cuttings being exhausted to the sea. It is possible to use a subsea diverter, which is deployed on top of the conductor on the seabed. If the well flows, the diverter bag preventer closes to stop the gas from being channeled up to the rig by the riser pipe and the gas exhausts to the seabed. Shear rams should be run as part of the diverter system so that the rig can easily disconnect the riser from the diverter and move off location.

Special Well Control Considerations

There are of course some circumstances where well control becomes quite complex. Some of these circumstances are briefly described to give familiarity with the terms and a basic understanding of what is meant.

High angle/horizontal well killing

The example discussed previously in this chapter assumed that the well was vertical. If the well is in fact drilled at a high angle, killing the well while maintaining a constant bottom hole pressure is a bit different. In this case, the driller's method of killing the well may be more appropriate. In this method, the well is killed in two circulations. No calculations are required for the first circulation; normal density mud is used to simply circulate out the influx. Pump pressure is held constant at PC_1 by manipulating the choke for the whole of the first circulation. Once the influx is out, if the pumps are stopped and the choke is closed, the pressure on both drillpipe and annulus should be equal to the original shut in drillpipe pressure.

For the second circulation, the only calculation required is the density of the kill mud, ρ_2. During Phase 1, the choke pressure is held constant at a pressure equal to the shut-in pressure at the end of the first circulation. Pump pressure will decrease during Phase 1 as heavy mud replaces light mud in the drillpipe. Once Phase 1 is complete, attention is switched to the pump pressure gauge and pump pressure is now held constant by manipulating the choke until heavy mud returns are seen from the annulus.

If a balanced kill is used, the Phase 1 graph is a bit more complicated to follow as it is not a straight line. This is illustrated in Figure 11-12. However, as Phase 1 proceeds, the pump pressure will tend to follow the line fairly closely with only small adjustments to the choke.

In a horizontal well, the cause of the kick is unlikely to be drilling into an overpressured zone. It may be a swabbed kick or some other cause. In this case, heavy mud is not required, it is just necessary to circulate out the influx while holding pressure over the chokes.

High pressure high temperature (HPHT) wells

As most of the shallow, easily exploitable oilfields have probably been found, the exploration for hydrocarbons has moved to remote areas and also to deeper prospects in mature areas. This has made HPHT wells much more common over the last few years. An HPHT well is defined as a well where wellhead pressure could exceed 10,000 psi and where the undisturbed bottom hole temperature exceeds 150° C. Though definitions in the industry do vary a little, this gives an idea of the scale of the problem.

From a well-control point of view, hole sections that meet the HPHT definition require extensive planning, careful training of drill crews and special equipment to detect and handle kicks.

Kick tolerance was mentioned in chapter 3. In HPHT wells, the formation fracture gradient and pore pressure gradient are often fairly close together, meaning the kick tolerance will be small. Therefore very sensitive

Drilling Technology in Nontechnical Language

Pump Pressure	1000	956	913	869	825	806	788	769	750	738	731
Strokes pumped	0	117	234	351	468	585	702	819	936	1053	1170
Barrels pumped	0	14	28	42	56	70	84	98	111	125	139

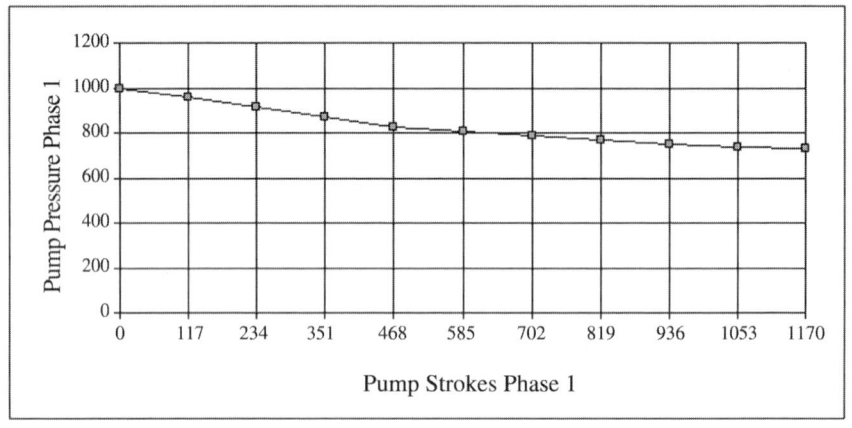

Fig. 11-12 Phase 1 Kill Graph for a High Angle Well

kick detection systems are necessary, together with crew training and regular drills to insure that at the first sign of a kick, the well is closed quickly (without visually checking for a flow) and the well is then watched for any buildup of pressure.

Another problem in HPHT wells is that oil-based muds are often preferred due to temperature limitations of some water-based systems. Gas at high pressure (above around 6,000 psi) is completely soluble in oil muds, so a gas influx can enter the well as a liquid. A small volume can enter undetected, but as this travels up the annulus and the pressure reduces (because the hydrostatic head of fluid above it reduces), the gas can suddenly come out of solution and expand to many times its liquid volume. The result would be a very rapid increase in flow detected at the surface and a large volume of gas to circulate out when killing the well. Under these circumstances, the kicking formation probably gave a continuous slow stream of dissolved gas into the mud, so that all of the mud between the kicking formation and the point where gas comes out of the solution is contaminated. More gas will come out of the solution during the well kill.

Special seals are required for the BOPs that can handle the high temperatures at which they have to operate. Also, as gas moves through the chokes it experiences rapid expansion due to the reduction from high pressure down to atmospheric pressure. As gas expands it cools down, and it is possible for the temperature downstream of the chokes to be well below freezing point. This requires consideration of low temperature steel strength and brittleness in the choke manifold. The normal working pressure of the choke manifold might have to be derated by 50% or more.

When gas in the presence of water cools down and reduces pressure, ice-like compounds called *hydrates* may be formed. Hydrates can plug lines during well killing, which presents a serious problem because it stops the killing operation. Hydrate formation can be prevented by injecting glycol into the choke manifold—so part of the preparation for HPHT drilling must include provision of an injection facility for glycol.

Relief wells

When a blowout occurs, the rig is likely to be damaged or even destroyed. It is also possible that the blowout will damage the wellhead in a way that makes it impossible to enter the well from the top in order to kill it. Under such circumstances, the only way to actually kill the well is to drill another wellbore to intercept the blowing well. The wellbore used to intercept the blowing well is called a *relief well*.

Once the relief well is close enough to the blowing well, the remaining rock in between can be fractured by pumping into the relief well at a sufficient pressure to create fractures between the wells. Once communication is established, heavy kill mud may be pumped down the relief well and into the blowing well.

One prerequisite to drilling a relief well is that the path of the blowing well is accurately known. Directional wells can be drilled with very high accuracy, but this is useless if the target position is unknown.

Underbalanced drilling

When a reservoir rock is penetrated by the drill bit, damage almost inevitably occurs at the exposed rock face. The major cause of this damage is interaction between the drilling fluid and the formation. The higher the overbalance, the greater the pressure tends to force mud into the formation and the greater the damage. If this overbalance can be eliminated, preventing most or all of the damage from drilling.

Normally, operations are conducted so that an overbalance always exists (primary well control is maintained). However, using special equipment, drilling fluids, and operating techniques, it is possible to drill with the mud hydrostatic pressure below the formation pore pressure. This means also that the formation fluids will continually flow into the well while drilling.

Drilling Technology in Nontechnical Language

A special tool is added on top of the BOP, called a *rotating control head*. This seals on the drillpipe while allowing rotation and movement of the drillstring. It is necessary to do this so that the actual pressure at the bottom of the hole (drilling fluid hydrostatic pressure + surface pressure) can be controlled (Fig. 11-13).

Horizontal oil wells are drilled in the Austin Chalk in Texas, which produce so much oil during drilling that the wells are paid for by the time they are finished!

Fig. 11-13 Rotating Control Head—Williams Series 7000. Photo courtesy of Williams Tool Company Inc, Arkansas.

Certification of Personnel for Well Control

Most government authorities worldwide now require supervisory staff on drilling operations to be demonstrably competent in basic well control techniques. This requirement is most often met by taking a course and passing a test, which incorporates theory training and testing and practical training and testing, using a simulator. The certificate is only valid for two years; regular training and re-testing has to be done.

Many personnel in current drilling situations will only encounter a few kicks throughout an entire career. As with any skill, if it is not regularly practiced it gets rusty with time. While having a current well control certificate does not guarantee an ability to handle a well control incident calmly and properly, it should demonstrate that the knowledge and ability exist.

Chapter Summary

This chapter described the three levels of well control (primary, secondary and tertiary). Well control equipment (BOPs and control systems, chokes, kick detection equipment) was covered in some detail before an example of a kick and killing the well was given, with the necessary calculations. Some special well control situations were described in sufficient detail to give a basic understanding of the terms used. Finally, the current requirements for certification of personnel were included.

Glossary

[1] **Active system.** On a rig, several tanks are used to hold mud that returns from the well, is treated by various items of solids-control equipment (as described in chapter 5) and is pumped back down the well. The mud travels through each tank in turn before going back to the pumps. All of this mud is termed the *surface active volume*. Other tanks may hold mud in storage, not as part of the active system—these are the *reserve tanks*.

[2] **Toolpusher and drilling supervisor.** The people involved in drilling operations (on the rig and in the office) and a typical organizational set-up are described in chapter 12.

3 **Boyle's Law** states that for a fixed mass of gas at a fixed temperature, pressure is inversely proportional to volume. If the pressure is halved, the volume must double.

4 ρ_2. The Greek letter ρ (rho) is used to signify a density gradient. ρ_1 refers to the original mud density, ρ_2 to the kill mud density.

Chapter 12

Managing Drilling Operations

Chapter Overview

A typical drilling operation can vary in cost between $20,000 per day for a small landrig in an established area to more than $150,000 a day for a modern offshore rig. Deepwater exploration wells can cost more than U.S. $40,000,000. How these activities are managed has a large impact on the return on investment (ROI) for the operator. Almost every drilling decision ultimately has an ROI basis; "what is the most cost effective way to proceed?"

This chapter will describe the traditional "chain of command" and will also examine some of the contractual models that are used in drilling companies. Cost estimating and tracking is very important and will be looked at in some detail. Logistics is a vital component in any drilling campaign and some of the considerations for logistics are included.

What happens if a major incident occurs is quite interesting ; this too is described.

Personnel Involved in Drilling Operations

Figure 12-1 is an example organization chart showing the typical command hierarchy in a traditional drilling operation. It's useful to understand the various titles, roles; and responsibilities of the various positions.

Oil company/operator

Operations manager. Generally responsible for all operational matters to do with drilling and production related activities; will most often take overall command of the incident response team in the event of a major incident such as an explosion, a blowout, etc.

Drilling manager. The top drilling position in the country in most oil companies. The drilling manager will report to the operations manager. Responsibilities include the day to day running of the drilling department, liaison with government authorities, overseeing the work of the senior drilling engineers, acting as the ultimate technical decision-maker for problems or disputes. In the event of a major incident, the drilling manager will join the incident response team and will be responsible for technical decisions to bring the incident under control. The drilling manager may deputize for the operations manager.

Senior drilling engineer (SDE). This position has different titles in different companies, but most people will understand the title even if it's different in their company. Responsibilities include: daily responsibility for one or more drilling rigs, well planning, daily reporting of drilling activities, supervising the work of more junior staff members, daily liaison with the drilling contractor and other service companies involved in the operation, and evaluating tenders for services.

Logistics coordinator. May report to the drilling manager or SDE. Responsible for all activities related to obtaining, storing, and shipping equipment and personnel to the wellsite and back. Coordinates land, sea, and air transports; supervises the storage yards;and monitors rental equipment to insure it is returned promptly once finished.

Drilling engineer (DE). Reports to the senior drilling engineer; responsible for technical work on drilling programs, technical evaluations, and engineering studies; and other work as directed by the senior drilling engineer.

Technical assistant. Works with the drilling engineer to help compile reports, locate information, filing, etc.

Drilling supervisor (DS). Often called the "company man"; the most senior representative of the operator at the wellsite. The DS is responsible for coordinating all drilling related activities, using the drilling program as guidance but making changes as necessary for a safe and efficient operation; instructs the rig crews via the toolpusher; ensures that safe practices are maintained; manages the various contractors on hire to the operator (e.g.,

Chapter 12 • Managing Drilling Operations

mud engineer, mud loggers, directional drillers, etc). Any operator personnel at the wellsite must report to and coordinate their work with the drilling supervisor, even if they are not "drilling" personnel (such as the wellsite geologist). The drilling supervisor discusses the ongoing program with the toolpusher and the two coordinate the workload. The drilling supervisor should not instruct the drill crews directly but through the toolpusher or night toolpusher. First, because the drill crews report to the toolpusher and not the drilling supervisor, and second, because this may create unsafe situations if the drilling supervisor and driller are both unaware of some other planned activity initiated by the toolpusher (such as maintenance work on certain equipment).

Night drilling supervisor. In some areas, it is usual to have an assistant to the drilling supervisor who will be "on shift" during the night (6 p.m. to 6 a.m. is usual). In other areas and in the past, this position does/did not exist. For critical operations during the night, the drilling supervisor will usually wish to be present but as long as things are proceeding according to plan, the night drilling supervisor will normally be responsible for overall supervision.

The night DS will often prepare the daily drilling report for the drilling supervisor to check and send in the morning. On most operations, the daily drilling report is transmitted to the operator's drilling office by 6:30 each day.

Wellsite drilling engineer/petroleum engineer (WSDE / WSPE). Often the two terms—drilling and petroleum engineer—are interchangeable. Sometimes there is not a WSDE or a WSPE on the rig. The night DS supports the drilling supervisor in the daily activities; the position is often a training role to gain practical experience.

Wellsite geologist. Monitors the work of the mud loggers (if present on the rig). Keeps a record of the geology encountered. Witnesses the work of the wireline loggers. Will sometimes determine the correct setting depth for casing strings.

Mud engineer. An employee of the drilling fluids contractor. Responsible for building and maintaining the mud system as per the program. Tests the mud properties frequently and reports the results to the drilling supervisor and to the mud contractor office onshore.

Drilling Technology in Nontechnical Language

Other contractor personnel. There are many operations that may require the services of other personnel, either employed by the operator or by service companies contracted to the operator. These personnel will report to the drilling supervisor and will have responsibility for their own equipment and area of expertise.

Drilling contractor. The drilling contractor owns the rig and employs the regular supervisors and crews working on the rig. The rigsite personnel will almost always work an equal time system—a week or a month on the rig followed by the same time off. This requires two sets of supervisors (toolpushers, night toolpushers, and a camp boss) and four complete crews, two of which will be on the rig at any time (drill crews [driller and below], radio operator, cooks, crane drivers, etc).

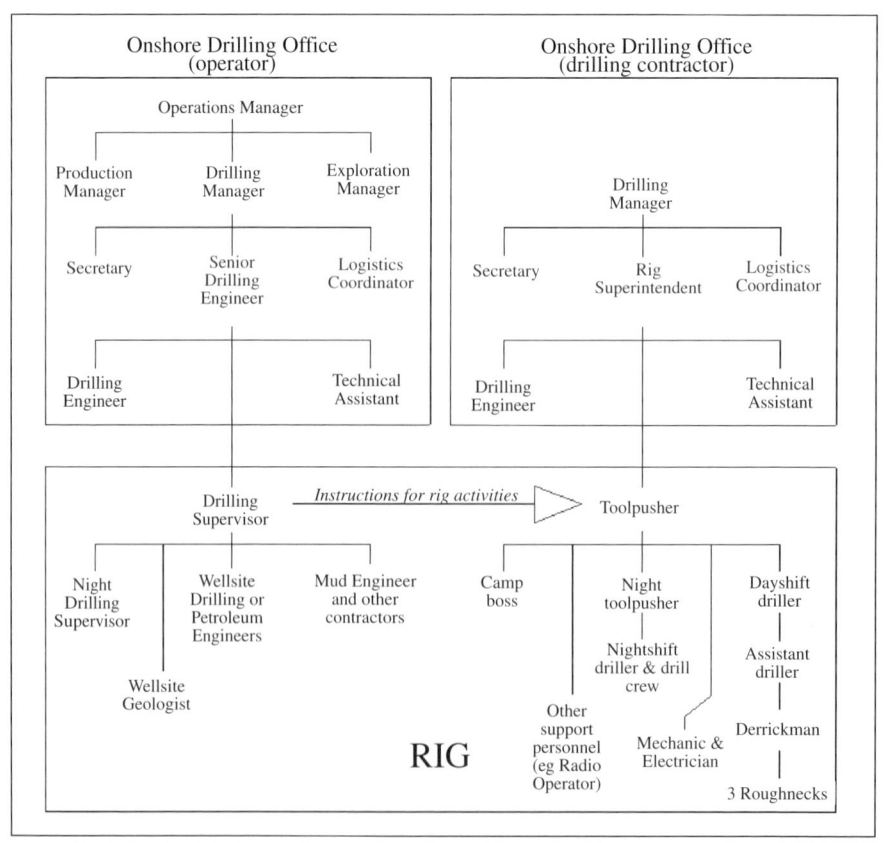

Fig. 12-1 Typical Drilling Operation Organization Chart

Chapter 12 • Managing Drilling Operations

Rig superintendent. Similar level of responsibility to the operator counterpart (senior drilling engineer); may be responsible for one or more rigs.

Toolpusher. The person in overall command of the rig. In some offshore areas (such as the North Sea) the toolpusher may also be designated as the offshore installation manager (OIM), with certain legal responsibilities similar to that of a captain at sea. (With floating rigs, whether anchored in position or not, there is often a qualified marine captain who is the OIM).

Night toolpusher. Usually works 6 p.m. to 6 a.m. Often deals with much of the paperwork and insures that the store is kept with sufficient spares, chemicals, etc. for upcoming operations.

Camp boss. In charge of accommodation and catering.

Radio operator. Usually two on the rig at any one time to provide 24 hour radio cover.

Driller. In charge of a crew of 5 or 6 people, the drillers usually work 12 hour shifts (most often from noon to midnight and midnight to noon) on the drill floor. The onshift driller is the most critical person on the rig. The decisions that the driller makes and how the driller reacts to problems has a huge influence on the final outcome.

Assistant driller (AD). Helps the driller by preparing tools and equipment, completes some of the paperwork, insures maintenance and repair work to drilling equipment is done (involving the mechanic or electrician as necessary).

Derrickman. Works up in the derrick when tripping into or out of the hole. Responsible for the mud pits and mud pump during drilling.

Roughnecks. Work on the drill floor as directed by the driller, AD, or derrickman.

Crane drivers. In charge of a crew of roustabouts working on the rig decks. Apart from operating the cranes, they handle other tasks around the rig (off the drill floor) as directed by the driller or toolpusher, and work 12 hour shifts to provide 24 hour cover.

Roustabouts. Wellsite laborers who work for the crane driver (or sometimes for a roustabout foreman). Normally people who work for a drilling contractor start their life at the wellsite as a roustabout.

Contract Types

There are now many different types of rig contracts that have evolved over the last 100 years. Lately, innovative contracts have been created that attempt to maximize the operator's return on investment by aligning the needs of the operator and contractors.

Two of the earliest types of contracts were footage and turnkey. These two types of contracts gave the contractor a pretty free hand in how the well was drilled, with little or no involvement by the operator.

In a footage contract, the drilling contractor was paid per foot of hole drilled and cased. The incentive to the contractor was to drill as fast as possible without any regard for the "quality" of the hole. Also there was a very strong incentive to take short cuts that exposed people to safety risks as well as potentially endangering the rig and the environment. Everything took second place to drilling as fast as possible.

A turnkey contract payed the contractor for completing the well and handing it over, ready to "turn it on"—hence the term turnkey. As with a footage contract, the incentive on the contractor was to drill and complete the well as fast as possible.

In these two contract models, serious damage to the reservoir is likely (or more accurately inevitable). The return on investment for the operator is low because the productivity of the well is low, due to the damage to the reservoir.

For many years, the most common contract model has been a day rate contract. It is still commonly used. The drilling contractor is paid a daily rate for the rig and has little or no say in how the well is drilled. The operator designs the well, writes the drilling program, and supervises the work at the wellsite. This allows the Operator control over the quality of the well, however the incentive to the contractor is to take as long as possible to drill the well!

In the mid to late 1980s, incentive contracts started to be commonly used. A day rate contract is still in place, but the contractor gains an incentive to organize the work well to finish the well faster. Now the objectives of the contractor and the operator start to be aligned—they share the same objectives. How well the objectives are aligned depends entirely on how the incentive portion is structured.

In a day rate contract, the operator usually contracts separately for all of the other services required. To drill a well might involve thirty different contracts for the provision of services ranging from aircraft to wellsite supervision. However, it's also possible that the drilling contractor provides some of these services under the rig contract, reducing the number of contracts and services that the operator has to manage.

One of the drawbacks to this day rate contract has been that each time a drilling campaign for exploration or development wells was planned, a complete set of tenders were issued that had to be evaluated. The evaluation in most cases was based primarily on cost comparisons, with little regard for the quality of services provided. This leads to below optimum performance, and in some cases (drilling fluids being a prime example), reduced production potential of the well, as well as a seriously reduced return on investment from what should have been possible. The more forward-looking operating companies such as BP and Shell now take a longer view, forming long term contractual relationships that can give greater benefits to both parties—a so-called "win-win" relationship.

Shell in the early 1990s developed a contracting strategy called *Drilling in the Nineties*, DITN. The basis of this strategy was that most of the Drilling activity could be contracted to a favored or "lead contractor" by contract strategies that rewarded the contractor for taking on board a large share of the responsibility for planning and drilling wells. One necessary feature of this is that contracts are long term.

From Shell's DITN strategy lead, other operators started to look into similar contracting strategies. One development of this was *Integrated Service Provider* (ISP) contracts, where one company with all of the required capabilities (some in-house, some subcontracted) would effectively manage complete projects. In the highest extreme an ISP might manage a field development project from initial surveying, exploration and development drilling, through production and final abandonment. Several major international contractors emerged with the capability to offer these complete services—Schlumberger, Halliburton, and Baker Hughes Inteq, being the largest. These multi-billion dollar groups compete for quite a large business pie, developing capabilities in drilling engineering by recruiting experienced drilling engineers and drilling supervisors (most of whom previously worked for the operators!), and by training the engineers and supervisors of the future.

There is one potential danger for the operators in this development. It's very important to their future business that they do not lose the capability to plan and supervise drilling themselves. If the operators lose this capability, they will also lose the capability to manage the risks involved in drilling wells. The major risks are those associated with a blowout, environmental damage, damage to the reservoir, etc. Whatever type of contract is in place, the operator cannot delegate those responsibilities. Unless a core of experienced people is retained, the operator must place its future business in the hands of the contractors.

Incentive Schemes

The purpose of an incentive scheme is to somehow change the way people work to meet particular objectives. There are various possible features to an incentive scheme, such as:

1. **Safety.** It is important that in meeting operational objectives, safety is not compromised. Some schemes have the typical unimaginative approach—something like "if anybody has an accident, everybody loses (part of) their bonus". This does *not* make people work safer, all it does is put a lot of pressure on people to not report accidents. It is, of course, inherently unfair on all those people who were not in any position to prevent or mitigate the accident, such as people off shift or working on a different part of the rig. Once earned bonus is taken away, the result is a positive disincentive to giving good performance.

2. **Lump sums for "flat time" operations.** A flat time operation is one where the depth of the well is not increased. The time-depth curve in Figure 12-2 was shown in chapter 3. The horizontal parts of the graph relate to operations such as running casing, or nippling up and testing BOPs. Instead of staying on day rate, the contractor could agree to complete some of these operations for a lump sum. This gives the contractor an incentive to plan and execute this part of the job efficiently and it limits the cost to the operator if the contractor has a problem or equipment failure.

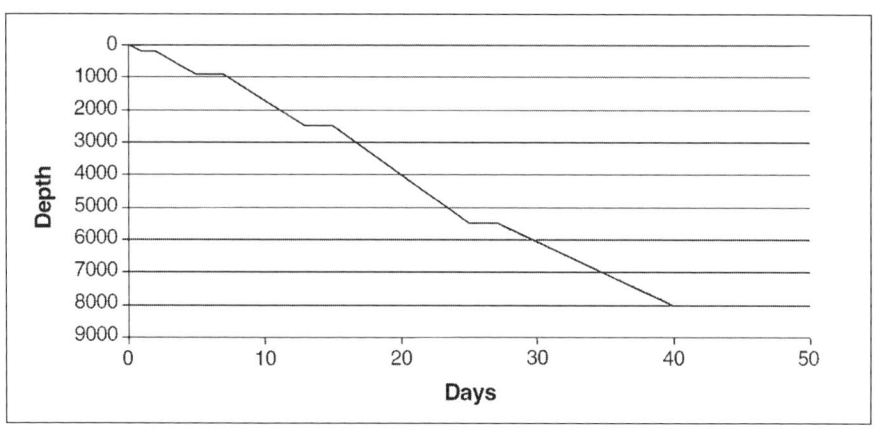

Fig. 12-2 Time—Depth for Well Example 1

3. **Meeting specific targets.** A target could be any performance related objective—cost at total depth, time to total depth, or anything that can be measured.

4. **Percentage of well cost saved.** The well could have an agreed target cost and for every dollar saved against that cost, the contractor earns a percentage.

For an incentive scheme to work well, it should include the following attributes:

- **It must be simple.** The people working for a bonus have to understand how to earn that bonus (what performance is required) and how much will be earned for any particular standard of performance. Some schemes are so complicated that it requires a computer spreadsheet to work out who gets what.

- **It must be easy to measure.** Ideally, anybody covered by the scheme should be able to read the daily reports or look at the time-depth curve (showing actual as well as predicted performance) and work out what's in it for him or her.

- **It must be fair.** If there are penalties for accidents then they should only apply to those people who were in a position to prevent or mitigate the accident. While bonus schemes generally reward the contractor company, the people doing the actual work should earn a fair share of the bonus.

- **It must be paid quickly.** If the bonus isn't paid for a long time then people become disenchanted with the scheme and with the operator. This acts as a disincentive.

Some schemes have a reward and a penalty, others have only reward elements.

Decision Making at the Wellsite

The division of responsibility between the drilling supervisor and the toolpusher must always be clear. This is vital not only for operational efficiency when making decisions, but also to insure safe working. With a day rate contract plus an incentive scheme, some operational decisions must pass to the toolpusher. The actual division will therefore depend on the contractual provisions.

A special case is decisions made during a major incident (one involving a risk of injury, damage to the rig, or to the environment). In some areas these responsibilities are prescribed by law (e.g., the OIM has legally defined responsibilities that cannot be delegated to another person).

These areas of responsibility must be defined before operations begin. All supervisory staff (whether operator, contractor, or service company) must understand and follow the procedures laid down.

On a typical day rate operation, the drilling supervisor will set out a daily program of work for the next 24 hours. This is then discussed with the toolpusher to insure that the contractor has some input and is able to flag any conflicts. For instance, if the drilling supervisor wishes to test the blowout preventer on a test stump[1] in readiness for nippling up the BOP later on, the toolpusher might be aware of some equipment problem that prevents the test from going ahead until later.

It is good practice for the main supervisory staff on the rig to meet at least once a day to discuss the upcoming program.

Decision Making in the Office

Each day, usually sometime around 6:30 a.m., the drilling supervisor transmits a daily drilling report back to the office. Previously this was by telex (and still is in some cases where communications are difficult) but now many operators use a computerized reporting system that builds a database as the reports are entered. Transmitting the daily reports will involve transferring computer files by e-mail or a direct file transfer method.

In a typical drilling office that has several rigs working, a morning meeting is held, chaired by the drilling manager. Each senior drilling engineer briefly describes the operations on his rig; any problems that have occurred or are expected and other relevant points. This is a good forum where others in the meeting may give advice. It also keeps everybody up-to-date on what everybody else is doing. People from other departments or sometimes from the drilling contractor or service companies might attend, though this is not usually the case.

Although the senior drilling engineer (SDE) responsible for a rig works office hours, the rig can contact the responsible SDE any time if problems occur. The SDE may offer advice to the DS or may take or confirm decisions.

The SDE also has to interface frequently with the other departments who have an interest in the well. The daily drilling report, either in full or an abbreviated version, is usually distributed to these other departments so as to keep them informed of progress and problems.

Interfacing with Service Companies

Unless an ISP is used to provide all of the various third party services, the SDE has to manage and coordinate the activities of up to 30 companies that may be involved in drilling a well. Much of the SDEs time will be taken up with these management activities.

Of particular importance are the vendors providing drilling fluid and cement for the well. If the same company is not providing both services, the SDE has to ensure that there are no incompatibility problems between mud and cement.

The service company representatives (salesmen) do provide a valuable service to the SDE. It is impossible for one person to keep up-to-date on all new developments in all service areas. By allowing time to meet with the representatives, the SDE can become aware of new tools, techniques, or services that may help to improve drilling performance. Often with new developments, the SDE can negotiate quite a good deal–as new developments need plenty of field applications before they become generally accepted. An SDE who is willing to try new things will return better overall performance than those who don't as long as the following simple rules are followed:

1. The new development has a reasonable chance of working, after a careful evaluation of the potential benefits and drawbacks
2. The cost of the development not working is not too high

For instance, say a drill bit manufacturer comes up with a new drill bit design. A deal can be made that guarantees the performance of the bit. The payment to the bit vendor is related to the bit performance. The bit vendor will want to insure that the bit is likely to succeed, not only because of the payment for the bit, but also because any failures in the field will make it harder to sell the bit to other companies. These deals can be quite attractive for both parties;if the bit works better than the anticipated performance, the well is drilled cheaper.

Estimating the Well Cost

Early on in the well planning process, the drilling engineer has to make an estimate of how much the well might cost. With a development well there will be a wealth of offset well data that will allow a reasonably accurate estimate, as long as the basic design elements of the well are known. Exploration wells are a different ball game because there is much less information available and many more unknowns. These unknowns are handled in the cost estimates as *contingencies*.

The reason for completing a cost estimate so early on in the process rather than after the drilling program is complete (when a proper accurate estimate can be made) is that the well has to be budgeted for in advance. A document known as an *Approval for Expenditure* (AFE) must be signed off by senior management.

A computer spreadsheet is an excellent tool for creating cost estimates. Once drilling starts, estimated costs can be replaced with actual costs, giving a continually updated forecast of the likely cost of the final well. In the event that the likely cost will exceed the AFE amount, an AFE amendment should be sought to cover the additional cost.

Well costs can be divided into several categories. Each cost element generally has a code that allows each to be tracked later on.

Fixed costs

Fixed costs are the same no matter how long the well takes to drill or how deep it is drilled. Typical fixed costs relate to moving the rig on location, moving it off after the well is complete, and surveying the well location.

Time dependent costs

A large proportion of the total cost of the well comes from the time it takes to drill the well. If problems occur, time-dependent costs escalate rapidly. On a day rate contract, the largest time-dependent cost will be the rig itself. Other time-dependent costs will include equipment on daily rental, personnel, vessels, helicopters, fuel, water, shore base, and dock fees. On an offshore well in deep water the rig might cost $100,000 a day, but the other time-dependent costs might bring the total daily operational cost to half as much.

Depth dependent costs

Depth dependent costs will increase as the well deepens. Typical depth dependent costs relate to casings, cement, completion tubings, drilling fluid, and drill bits. These might form up to a third of the total well cost, unless problems occur that substantially increase the time dependent costs.

Support costs (overheads)

Overheads are the costs that are incurred by the office and other off-rig activities. Examples include engineering work (such as data analysis or studies), support staff (such as secretaries) as well as a proportion of the cost of the office. Some of these costs are time dependent and some are fixed costs (Fig. 12-3).

Contingency costs

There are some problems that can be expected to occur, with a small or large probability that any particular problem will actually occur. The cost of each event—the contingency cost—is the probability of its occurrence multiplied by the cost if it actually does occur. For instance, in a particular hole section, offset data predicts that mud losses might occur. If the cost of those losses might be $50,000 and the likelihood of losses is 10% then the contingency cost is $5,000 (Fig. 12-4).

Accuracy of the estimate

There is one other element to the cost estimate. The accuracy of the estimate—the confidence in the final figure—must be given. If good offset data is available, the accuracy generally will be that the final cost should be within 10% of the estimate (exceptionally it is within 5%). For an explo-

Drilling Technology in Nontechnical Language

Well cost estimate - level 2

Well: Exploration Well 1 **Date:** 18-Jun-99
Reporter: Steve Devereux

A/C no.	Description	Cost rates Rig Move	Drill/Suspd	Compl/Test	Cost Estimate Rig Moves 3.0 days	Drill/Suspd 29.9 days 2900 m	Compl/Test 8.0 days 2900 m	Total 40.9 days
TIME DEPENDENT ($/day)								
87201	Rig Rate	24,000	25,000	25,000	72,000	748,688	200,000	1,020,688
87415	Vessels	7,000	7,000	7,000	21,000	209,633	56,000	286,633
87230	Additional (catering etc)		400	600		11,979	4,800	16,779
87530	Cement serv. & pers.		508	508		15,213	4,064	19,277
87524	Mud logging		600			17,969		17,969
	Conductor driving equipmt		2,000			10,000		10,000
87665	Dock fees & base overheads	4,400	4,400	4,400	13,200	131,769		144,969
87551	Rental tools		700			20,963		20,963
87554	Consultants on rig	1,400	1,400	1,400	4,200	41,927		46,127
87551	Anderdrift survey tool to 2450m		500			5,700		5,700
87506	ROV mob; drill 26", set 20"	2,100	2,100	2,100	6,300	4,201		10,501
87554	Water	5	5	5	15	150	40	205
87554	Fuel (incl rig and vessels)	500	500	500	1,500	14,974	4,000	20,474
	TOTAL		45,113		118,215	1,233,164	268,904	1,620,283
DEPTH DEPENDENT ($/m)								
87545	Deviation survey (gyros, MMS)		5			14,500		14,500
87315	Mud and chemicals					226,509		226,509
87330	Solids control consumables		2			5,800		5,800
87320	Cement and chemicals					25,924	1,620	27,544
87301	Bits					133,920		133,920
86201	Casing and accessories					889,610	46,005	935,615
86201	Completion			30			87,000	87,000
	TOTAL					1,296,263	134,625	1,430,888
FIXED COSTS ($)								
87110	Site survey	100,000			100,000			100,000
87220	Rig positioning	25,000			25,000			25,000
87220	Rig Mob/Demob							
87220	Boats Mob/Demob	48,000			48,000			48,000
87548	Casing crews & equipment		20,000	5,000		20,000	5,000	25,000
87509	Electric logging		498,886			498,886		498,886
87509	Cased hole logging and perf.		17,498	77,382		17,498	77,382	94,880
87554	Well Testing			100,000			100,000	100,000
86211	Wellhead		50,000			50,000		50,000
87554	Insurance	20,000			20,000			20,000
	Fishing & Abandon services		10,250			10,250		10,250
	Well planning		40,000			40,000		40,000
	TOTAL				193,000	636,634	182,382	1,012,016
SUPPORT COSTS ($/day)								
87554	Drilling Office overhead	2,500	2,500	2,500	7,500	74,869	20,000	102,369
87665	Office Sup't consultant	1,000	1,000	1,000	3,000	29,948	8,000	40,948
87554	Other drilling expenses	50	50	50	150	1,497	400	2,047
87420	Air transport	5,000	5,000	5,000	15,000	149,738	40,000	204,738
	TOTAL				25,650	256,051	68,400	350,101
	GENERAL TOTAL	8,550	8,550	8,550	336,865	3,422,112	654,311	4,413,288
	TOTAL							**$4,413,288**

Fig. 12-3 Cost Elements by Code

Chapter 12 • Managing Drilling Operations

Well cost estimate - level 2						
Well:	Exploration Well 1					
Reporter:	Steve Devereux		Date:	18-Jun-99		
			Total cost of all events =		$	1,602,564
Contingencies			Total contingency cost =		$	586,630
Problem	Probability, %	Rig days	Other costs	Total cost	Contingency cost	
Problems running 20" casing into hole	25	2		$ 107,326	$ 26,832	
Hole instability in 17 1/2" hole	25	5	$	$ 268,315	$ 67,079	
Hole instability in 12 1/4" hole	25	5	$	$ 268,315	$ 67,079	
Kick in 12 1/4" section	25	2	$	$ 107,326	$ 26,832	
Kick in 8 1/2" section	25	2	$	$ 107,326	$ 26,832	
Losses in 6" section	50	3	$	$ 160,989	$ 80,495	
Set 5" liner & TD in 4 1/2"	50	7	$ 100,000	$ 475,641	$ 237,821	
Logging problems	50	2		$ 107,326	$ 53,663	

Fig. 12-4 Contingency Costs

ration well, the accuracy will be considerably less—25% would be a reasonable assumption.

For the total estimated cost, all of the previously mentioned categories are added up. This is considered to be the likely cost. To arrive at the maximum cost, the total is multiplied by the accuracy factor. For a total of $5,000,000 and an accuracy of 10%, the maximum cost of the well should not exceed $5,500,000. This is the amount that the AFE should approve (Fig. 12-5).

Logistics

Logistics is the art and science of getting people and materials to the rig and back again. Without effective logistics, the well will not get drilled. Logistical problems increase with the distance from the point of supply and the remoteness of the location. Some special situations (such as in remote mountainous or heavily forested areas) might mandate the use of a rig that is totally moved and supplied by helicopter.

As well as moving supplies, their whereabouts must be monitored. Many drilling operations if not all now use computerized inventory and monitoring systems that allow close tracking to be maintained.

Drilling Technology in Nontechnical Language

Well cost estimate – level 2
By: Steve Devereux

BUDGET TITLE	Exploration Well 1		**Notes** 1. Assumes well logged, tested and abandoned.
			2. Mob/Demob costs **are not included**.
DATE	18-Jun-99		

VALUE of this estimate	$4,999,918	
Dry Hole cost of well	$4,345,607	($654,311 less than total)
Contingency included	$ 586,630	(Equivalent to 12% of the total estimate)
Accuracy of estimate	15%	(Maximum anticipated cost $5,749,906)

1) SUMMARY

This well cost estimate covers the drilling, logging, testing and abandonment of the Exploration well.

2) ESTIMATES

Well time estimate:

Rig move, jack up, preload	at 50 m	2.0 days
Drive 30" conductor	at 200 m	2.0 days
NU diverter		1.0 days
Drill 26" hole	at 1000 m	2.0 days
20" surface casing		1.2 days
17.5" hole	at 1700 m	2.0 days
13 3/8" casing		1.2 days
12.25" hole	at 2450 m	4.0 days
9 5/8" casing		1.2 days
8 1/2" hole	at 2900 m	4.0 days
7" liner		2.0 days
TD logging		3.1 days
Testing		6.0 days
Abandon well		3.0 days
Release rig		1.0 days
Weather downtime 5%		1.7 days
Rig downtime 10%		3.6 days
TOTAL:		**40.9 days**

Well cost estimate (see attached for details):

Rig move:	$ 336,865
Drill, Log, Suspend:	$ 3,422,112
Compl & Test:	$ 654,311
TOTAL Base Estimate	**$4,413,288**
Contingency	$ 586,630
TOTAL with contingency	**$4,999,918**

Fig. 12-5 Cost Estimate Summary

Some materials have restrictions as to how they can be moved. Explosives; toxic, corrosive, or flammable chemicals; and radioactive materials all require special precautions and may not be transported by helicopter.

Each chemical is accompanied by a form when it is transported. This form is known as SHOC—Safe Handling of Chemicals. The SHOC form identifies the chemical and it's dangerous properties, lists any special handling or storage procedures, and states what first aid procedures to use if somebody is exposed to it.

One of the early considerations in a drilling campaign has to be whether the infrastructure exists to supply the rig. This may mean building roads, heliports, harbor facilities, and storage areas. As these things take time, any deficiencies in the infrastructure must be known about early enough to remedy them.

Handling Major Incidents

Before operations commence, there must be an Incident Response Team in place. The role of this team is to assemble in a predefined location in the event a major emergency takes place, such as a fire or explosion, blowout, or structural failure of the rig. The team members and the specific role of each is defined.

All responsible operators and drilling contractors have a room set aside that can be rapidly set up with all of the necessary facilities for the team to function. These facilities will include communications (telephone, telex, fax, radio, and e-mail); a log board that is used to keep a visible log of events, diagrams of the rig and other facilities, and contact details for all the emergency services and other operators who may be able to assist. There will be sufficient desks and chairs for the whole team to work, plus representatives of the emergency services who might send a representative along to coordinate their services directly with the team (such as police, coastguard, or fire brigade).

Training of team members is important. If the team is to function well together then they must train together for the event. There are companies that specialize in putting together realistic simulations of major events, and it's surprising how much pressure can be placed on the team members in these simulations! It's good if a couple of days a year can be put aside for such training.

When a major incident occurs, the rig supervisor (drilling supervisor, toolpusher or OIM) will assume the position of onscene commander and initiate a procedure to call out the response team. The team will assemble in the incident room and the first person there will start the log board, noting times and events. As more team members arrive, they will be able to see the log board and get a quick view of the situation. It's very important that the team members allow the onscene commander to get on with handling the incident at the wellsite by giving support and coordination of resources, rather than by trying to pull the strings from onshore (which they don't have the legal right to do). The people on the rig are (or should be) qualified and competent to react properly to the situation, and they are certainly in the best position to assess what's happening. Here is also where legal responsibility lies. However, the experienced operations staff present in the incident room will be able to look at the data coming in and advise the rig on alternative scenarios that, in the heat of the incident, the wellsite staff might have overlooked.

The response team will communicate with the emergency services, government authorities and the media. Handling the media is an important consideration and it's much better if one person is assigned to keeping the media briefed without releasing names or unconfirmed data. Once the media sniff an incident (and they'll find out about it surprisingly quickly) they will be all over the place, trying to get any small detail. Information such as names or numbers of casualties must never be given over an open radio channel as these will be monitored by the media and others. If casualties are involved then the police will take over the task of informing relatives as well as gathering information for later investigation.

Once the incident is under control, the investigation of the causes and chain of events begins. The police will want to visit the scene of any casualties (serious injuries or fatalities). They generally have the power to demand that the scene not be disturbed (more than is necessary to secure the rig and prevent further damage or problems). If major pollution has resulted then the cleanup might take some time and cost a lot. The aftermath can take a while to get through. As much as possible must be learned to reduce the chances of any recurrence.

Chapter Summary

This chapter looked at a variety of subjects relating to the management of drilling operations. Job titles and responsibilities, contracts, incentive schemes, and cost estimating were covered. Logistics and responding to a major incident were described.

Glossary

[1] **Test stump.** A "false" wellhead spool to which the BOP can be connected and tested. Once the BOP is then nippled up on the actual wellhead, only the last connection needs to be tested. This can save a considerable amount of rig time.

Chapter 13

Drilling Problems and Solutions

Chapter Overview

There are various common problems that might occur during drilling and plenty of uncommon ones too! This chapter will describe the most commonly encountered drilling problems caused by downhole conditions (as opposed to surface equipment failure).

When any problem is encountered, the drilling staff must identify the root causes of the problem before a proper response can be formulated. If the root causes are incorrectly identified then it's unlikely that the correct response will be made, and this might not cure the problem. It can even make the situation worse. Identifying the root causes of problems covered in this chapter will be described along with possible responses.

Lost Circulation

As mentioned in chapter 3, lost circulation occurs in varying degrees.

Losses of up to 30 barrels an hour would be called *seepage* losses. This condition is usually caused by drilling through very permeable formations where the mud cannot form an effective filter cake.

Between 30 and 60 barrels of mud lost per hour would be *moderate* losses. Again, this is likely to be caused by high permeability formations and an ineffective filter cake. It could also be caused by faults that do not seal, but allow mud to enter into the fault system.

More than 60 barrels an hour of mud lost downhole are *serious* losses. This level of losses is unlikely to be caused by high permeability formations. Potential causes include non-sealing faults or fracture systems.

If the mud losses are so severe that no returns are seen at the surface, the term *total* losses would be used to describe the situation. Potential causes include non-sealing faults or fracture systems and drilling into formations that contain large caverns (also called *vugs*).

Losses occur because the following conditions are both present:

- The drilling fluid overbalances the problem formation
- There is a path that allows the mud to flow into the formation and away from the wellbore

Preventing and curing losses addresses these two factors—reduce overbalance and plug off the pathways.

In discussing lost circulation, it is convenient to categorize the problem according to situations in which the losses occur.

Losses while drilling surface hole

Losses in surface hole have two common causes:

- Very permeable formations (often unconsolidated sands) that allow whole mud to seep through the pore spaces. Commonly seepage to moderate losses might be caused by this. Total losses into permeable formations are unlikely.
- Fractures (either existing or more likely created by the drilling process) that allow mud to leave the wellbore. Severe or total losses are likely with fractures.

Losses into a permeable formation might be cured by pumping lost circulation material (LCM) into the well. LCM is bulky, lightweight plugging material that blocks the pores in the formation. Materials used for LCM include clay, sawdust, mica (a stable non-hydrateable clay mineral with large, flat, plate-like crystals), and ground nut shells. Usually some mud would be placed in a separate tank—about 100 bbls—and coarse LCM mixed in. The drill bit will be pulled up the well until it's somewhere close

Chapter 13 • Drilling Problems and Solutions

above the top of the permeable formation, the LCM mud is pumped to the zone and left for a while. It is hoped that the LCM will create a nice plaster of material at the exposed face of the formation. Sometimes it works and sometimes it doesn't.

As well as attempting to plug the formation face, the pressure exerted on the formation must be minimized. There are several techniques that are available to achieve this:

1. Reduce the density of the drilling fluid itself. *Spud mud* commonly used to drill surface hole is usually a simple mix of bentonite (which has a low specific gravity) and water. The bentonite disperses in the mud and the individual clay crystals, which are plate-like, can form a good filter cake on permeable formations—unless the formation is extremely coarse. Bentonite also imparts viscosity to the mud. This is called unweighted mud because no weighting materials (such as barite) are added.

The density could be further reduced by:

   a. mixing the mud with fresh water rather than sea water, if the supply situation permits.
   b. aerating the mud using compressed nitrogen injected at the standpipe (Aerated mud is discussed in chapter 7).

2. Reduce the density of the drilling fluid when it is contaminated with cuttings, which it will be while drilling in the annulus.

Several of the following methods may be easily employed:

   a. increase the viscosity of the mud at the shear rates present in the annulus so that the mud lifts out cuttings more efficiently. Polymers (such as starches) can be added to do this.
   b. drill more slowly so that cuttings contaminate the mud at a lower rate. This is called *controlled* drilling because the ROP is controlled at a level that the bit could potentially drill below if the drilling parameters were optimized.

c. increase the circulation rate. This lifts out cuttings more quickly. It will also increase the annular pressure losses slightly, but this effect is very small due to the large diameter, shallow hole typically drilled for surface casing. In most cases, the reduction in contaminated mud density causes a greater drop in bottom hole pressure than the increased annular pressure loss.
d. from time to time, pump around a pill of highly viscous mud to try to clean out any cuttings not moving out of the well.

3. Reduce the height of the mud column. On an offshore rig, a surface hole might be drilled with returns to the seabed, rather than trying to create a closed circulating system. It's also possible to weld a remotely-controlled large-diameter valve on the riser close to the sea surface. If losses are taken, the valve can be opened and returns exit at sea level. If the well flows, this valve is closed at the same time as the diverter.

While using these techniques, maintaining primary well control and monitoring for potential kicks must not be compromised. The risks are different for an exploration well than for a development well.

Fractures are a real cause for worry. If the fracture extends to the surface, the rig itself might collapse if the ground underneath it loses strength due to these fractures and the lubricating effect of mud in them. These surface hole losses *must be cured* and the only effective way is by pumping cement into the well and insuring that the cement sets in the fractures close to the wellbore. The procedure for placing cement in the right location while it sets is described later.

If a pilot hole is drilled and later opened up, the loss situation can be controlled much more easily. There is more available annular velocity in the pilot hole at the maximum pump output rate due to the smaller hole capacity and the fact that the physical area of the wellbore wall is smaller (therefore less area for mud to escape). When the hole is opened, the ROP is controlled at a rate that prevents excess loading in the annulus by cuttings.

On platform wells, if all of the conductors are set before drilling begins, it's possible for losses to become established between the well being drilled and an adjacent conductor. The distance through the rock can be very short (a few meters). The other very real danger here is that if shallow gas hits, gas might flow via an adjacent conductor that forms a perfect conduit to the surface and without any diverter set. In this situation, conductor depths may be staggered to increase the distance between conductor shoes. It is also very good practice to set around 50' of cement in the bottom of each conductor before starting to drill the first well.

Losses in normally pressured, deeper formations

These formations may be unconsolidated, naturally fractured, become fractured by the drilling operation, or consolidated but highly permeable with pore sizes too large for the mud solids to plaster. The loss zone can be anywhere in the open hole—not necessarily the formation just drilled into.

Several factors can contribute to the mud loss, such as the annulus becoming loaded with cuttings, high ECD, excessive mud density, insufficient mud viscosity, high water loss (low solids content to plaster the wall), excessive surge pressures, breaking the formation during an FIT, or closing in the well after a kick.

It is necessary to identify the type of loss zone and the mechanism causing the loss. Knowing the depth and type of loss zone will help formulate a strategy to cure the losses.

In general, several techniques can be useful in most loss situations deeper in the smaller-diameter wells. In order of attempting, these are to:

1. Decrease circulation rate for lower ECD and drill with controlled parameters to minimize annulus loading. The losses may well decrease over a period of time (a few minutes to a few hours). The severity of the loss may dictate whether this is acceptable; for instance, with losses of more than 60 bbl/hour using expensive OBM the cost may be too high. Restricted replacement mud supply may also preclude this.

2. Reduce mud density if possible by diluting the mud and maximizing the use of solids-control equipment.

3. Pump a 100 bbl LCM pill with mixed fine, medium, and coarse LCM. Place it across and above the loss zone and observe the well. When the well becomes static, start circulation cautiously, monitoring the active volume, and resume drilling.

4. Add solids (LCM) to the whole active volume of mud to increase plastering characteristics

5. Severe or total losses can sometimes be cured by drilling ahead slowly with reduced weight on the bit. It is only an option if there is a plentiful supply of water (such as on an offshore rig) and mud chemicals.

Drilling with no returns is called *blind* drilling. It should not be attempted if there is any chance of hydrocarbons being encountered. The circulation rate must be sufficiently high to lift cuttings away from the BHA and up to the loss zone, where they may plug off the large pore spaces or fractures. A general lower limit for annular velocity for blind drilling is 50 feet per minute.

Further action may include setting a barite plug (described in chapter 11), diesel oil bentonite plugs or cement plugs (described later in this chapter), setting an extra casing string, or plugging back and side-tracking if the severity of the situation warrants it. Aerated mud techniques described in chapter 7 might also be applicable, but these must be planned well in advance.

Losses in heavily fractured or cavernous formations

Some formations contain very large fractures (typical of limestone reservoirs) or large caverns. When the bit penetrates one of these spaces, total losses are likely to start immediately—a good sign that this type of formation has been penetrated.

One option may be to drill blind with water if the area is well known or to drill with foam if not. Set casing as soon as possible after drilling right through the loss zone (drilling far enough to give a reasonable column of cement above the shoe).

Another option is the careful spotting of large volumes of cement as described later, which may or may not work. Several successive attempts may be needed to progressively plug off fractures or caverns around the wellbore.

Diesel oil bentonite plug

DOB plugs are also known as gunk plugs. They work by holding bentonite in suspension in diesel until the plug is in place downhole and then arranging for water to hydrate the bentonite. The Bentonite hydrates rapidly, becoming extremely viscous. They can be very successful in shutting off flow in an underground blowout, especially if the flow is water.

A DOB plug will not maintain strength indefinitely. Cement should be spotted to give a permanent seal once the plug has worked. The main problem with the DOB plug is that it will set up inside the drillstring if it contacts any water. A good diesel spacer ahead and behind the plug are essential to prevent this.

To place the plug, 150 lbs of bentonite is mixed for each barrel of diesel. Normally a DOB plug will be between 30 and 150 barrels in size. There must be no water present in the mixing pit or in the pump lines. About 10 barrels of diesel is pumped ahead as a spacer, followed by the DOB slurry, and then another 10 barrels of diesel behind the slurry. Fresh water is pumped behind the diesel. The plug is pumped down the drillstring to the loss zone. As the water behind exits the drill bit, the larger annular capacity allows the water to mix with the DOB slurry. Any water already present in the loss zone will also mix with the slurry, hydrating the bentonite. Water can be pumped ahead of the first diesel spacer if desired, in which case the diesel spacer ahead might be increased in volume.

Curing total losses with cement

The best lost circulation material for severe to total losses is cement. The following are two keys to success that the procedure must aim for:

1. Sufficient cement must be placed in the loss zone immediately around the wellbore

2. The cement must not move away from the near-wellbore zone while it sets

Many times a small volume of cement is used and the subsequent actions almost guarantee the cement will move away from the wellbore after placing. The following is an outline procedure that has worked well in the field:

1. Drill right through the loss zone so that it's completely penetrated, if possible.

2. Position drillpipe just above the top of the loss zone.

3. Pump a large quantity—a couple hundred barrels—of lightweight extended cement slurry

4. Pump a large quantity—a hundred barrels or so–of neat cement, which incorporates polypropylene fibers in the slurry. The slurry should be designed so that the compressive strength when set will be less than the formation compressive strength. This is to avoid the drill bit from drilling an unintended sidetrack. If the fibers plug off against the formation face, that's fine—it will prevent the earlier slurry from moving away from the wellbore.

5. Pump mud behind the cement. The volume of mud pumped should be some barrels less than the total drillpipe capacity so that a bit of cement is left in the pipe—about 5 barrels.

6. Pull out a couple hundred feet of the drillpipe without putting any mud into the hole. Normally, mud is pumped into the annulus when the pipe is pulled out to insure primary well control is maintained (described in chapter 11).

7. Pull out another few hundred feet of pipe. Pump a measured amount of fluid into the annulus while pulling out; no more than is needed to replace the volume of steel removed. It is better to pump too little than too much so as to avoid chasing cement away from the near-wellbore zone

8. Wait until the cement should be hard, then top the well with mud. Pull out the drillpipe, run in with a drilling assembly and drill out the cement

While the procedure will vary a little depending on the actual situation, this procedure will give a high probability of success the first time.

Stuck Pipe

The drillstring can be said to be *stuck* when it cannot be pulled all the way out of the hole without exceeding the maximum allowable pull on the pipe. Within that definition, it may or may not be possible to circulate—pipe movement downwards below the stuck point may or may not be possible and pipe rotation also might or might not be possible.

Most cases of stuck pipe are avoidable with good supervision. Pipe rarely gets stuck without advance warning signs. If the conditions are present that make stuck pipe possible then the drilling supervisor and drillers must stay on the ball and take suitable precautions to prevent a stuck pipe from occurring.

If the pipe does get stuck in such a way that no pipe movement is possible, the actions taken in the first few minutes will usually determine the outcome. Sticking forces increase with time so it's very important to analyze the situation and take immediate action to get the pipe unstuck.

It is convenient to classify stuck pipe into categories of the main cause of sticking.

Sticking mechanisms for stuck pipe

Causes of stuck pipe can be classified into the following basic categories:

1. **Geometry** is related to dimensional problems in the wellbore. Circulation is usually possible, the problem will be seen with the string moving in only one direction.

2. **Solids** is related to solid particles in the hole. Circulation may be restricted or impossible and hole cleaning may have been inadequate. Usually occurs when pulling out of the hole.

3. **Differential** sticking is related to mud overbalance over a permeable formation.

Each of these will be examined in turn.

Geometry related stuck pipe

There are several different root causes of geometry related stuck pipe.

If the wellbore is *undergauge* for any reason, then full gauge tools such as the drill bit or stabilizers can get stuck if moved into the undergauge part of the hole. Several causes are possible for undergauge hole—drilling with an undergauge bit, thick filtercake buildup on a permeable formation, mobile formations (squeezing salt or shale that moves into the wellbore with time).

In deviated wells, it is possible for a *keyseat* to form. Where the drillstring (OD of 5") presses against the inside of the bend in the wellbore, it can wear a groove into the formation. If this groove becomes deep enough, when the pipe is pulled out of the hole the drillpipe will slide inside the groove. However, the top of the BHA cannot move through the groove, having a larger OD. The result is that the pipe becomes stuck at the top of the BHA. Full circulation will be possible. If the BHA was not pulled too hard into the keyseat it might be possible to move the drillstring down.

Keyseats can be avoided by limiting the dogleg severity of the build up section—limiting the number of rotating hours against the bend before it is cased and limiting the drillstring tension while drilling below the bend. A tool known as a string reamer can be placed in the drillpipe above the bend, this tool enlarges the groove as it moves downward during drilling. A keyseat reamer can also be run at the top of the BHA. This allows the keyseat to be dressed out by rotating while pulling out of the hole. It takes extra time and is a bit tricky so it's preferable to use a string reamer if possible (Fig. 13-1).

If a section of the hole was drilled with an assembly that caused a dogleg to form (such as a build or steerable motor assembly), problems can occur if a more highly stabilized assembly is run straight into the same section. It's a geometry related problem. It is necessary to ream first with the new BHA. Reaming is like drilling, but instead of rotating on bottom, the bit drills on its gauge area only to smooth out the curved section and enlarge it slightly. The work that the bit does is monitored by watching—and limiting—the torque required to turn the drillstring.

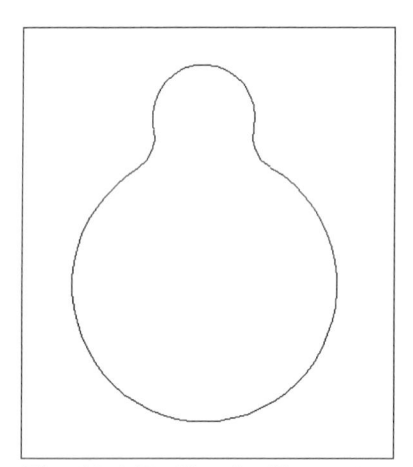

Fig. 13-1 Profile of a Keyseat

Where hard and soft layers of formation alternate, the soft parts can enlarge more than the harder formations. *Ledges* form. It is possible for the pipe to get stuck at changes of diameter (such as on the bit) on stabilizers, or at the top of the BHA.

Solids related stuck pipe

Solids particles in the annulus can cause the pipe to get stuck. Mostly these solid particles will be drilled cuttings or wellbore cavings. However, there are other possibilities. Solids related problems normally occur when pulling pipe out of the hole, or if the pipe is left in one place without any circulation. Almost always, circulation will not be possible because the solid particles will block off the annulus. If circulation is impossible due to solids, the hole is said to be *packed off*. This stuck pipe situation is the most difficult to cure, but it's luckily not to difficult to avoid, most of the time.

Before the drillstring is pulled out after drilling, circulation should continue after drilling stops long enough to lift all the cuttings to the surface. This is not always as straightforward as it sounds. In inclined wellbore sections, cuttings move upward by a combination of lifting (vertically upward movement) and rolling along the low side of the hole. As inclination increases, more rolling and less lifting takes place; in a horizontal well, the vertical lift component is zero. Making solids roll along the hole takes more energy (higher flow rates) than lifting them. Also, as inclination increases, the drillpipe will tend to lie on the low side of the hole. This reduces the flow rate along the bottom, as the flow will preferentially go to the largest available area—above the drillpipe. Under these conditions, cuttings beds form easily and are not easily moved.

Stable cuttings beds can form in inclinations more than 60°, especially if the hole is overgauge in places. At lower inclinations, any cuttings bed will reach a certain size and then avalanche down the hole. Stable cuttings beds sit there nice and quietly until the BHA is tripped out of the hole. As the larger BHA encounters the bed, it ploughs into the pile of solids and can get stuck. If the pipe is packed off with solids then the prognosis is bad. To cure the problem, circulation must be re-established, but this is often impossible.

Another source of solids in the well is *reactive formations*. This is overwhelmingly a problem with shales. Formations can become unstable due to adverse reactions with the mud physical or chemical properties. Two instability modes are possible. The shale hydrates, becomes plastic and sticky, and falls into the wellbore. Otherwise the hydrostatic pressure exerted by the mud is insufficient to hold the formation together and slivers of shale fall off into the well as cavings (mentioned in chapter 3). The same can happen when drilling in overpressured shales—the pore pressure might even be higher than the mud hydrostatic, but the well doesn't kick due to the very low permeability of shale and high levels of cavings result. Even if the annulus is clean of cuttings at the start of a trip out of the hole, reactive formations can cause material to fall in during the trip and cause the pipe to stick. The warning signs are that as the steel pipe is removed from the hole the force needed to pull the drillstring out does not decrease as rapidly as would be expected. If this occurs the driller must stop pulling out and circulate the hole clean again.

In shallow unconsolidated sands, *tophole collapse* might occur. The hydrostatic pressure must be able to support the formation, but if there isn't a good filter cake and losses occur, there is little difference in pressure between the formation around the wellbore and mud hydrostatic. The sand simply falls into the well. Try digging a deep hole on a beach! The mud must have good plastering characteristics so a firm filter cake is formed, allowing the mud hydrostatic to hold the sand back.

Sometimes, formations are *fractured*. This can happen especially with brittle shales, coal, and limestone. Mud gets into the fractures, lubricating the fracture faces as well as changing the pressure regime near to the wellbore. Pieces of formation fall off (fracture cavings). Cavings from fractured formations can be identified by their size, shape, and presence of fracture faces that may be visible.

Junk in the hole can also fall into this category. If something is lost down the hole, either because of something breaking or by something falling through the rotary table, stuck pipe may result.

Cement that is not fully set is known as *green cement*. If a cement plug was set in the hole and the driller runs into it with the bit while it is still "green", the pipe can become plugged or can get stuck in the cement.

Differentially stuck pipe

Differentially stuck pipe is far less common today than 15 or 20 years ago. The knowledge of the mechanisms involved, better mud, more use of stabilized BHA's, and better awareness of drill crews has relegated differential sticking to a minority of cases of stuck pipe. Interestingly, many people immediately assume differential sticking when a problem occurs and have to be convinced that something else might be the cause (Fig. 13-2).

Four conditions are identified that must all be present for differential sticking to occur. These are the following:

- The presence of a permeable zone covered with wallcake
- Static overbalance on the formation

- Contact between the wall and drillstring
- A stationary string

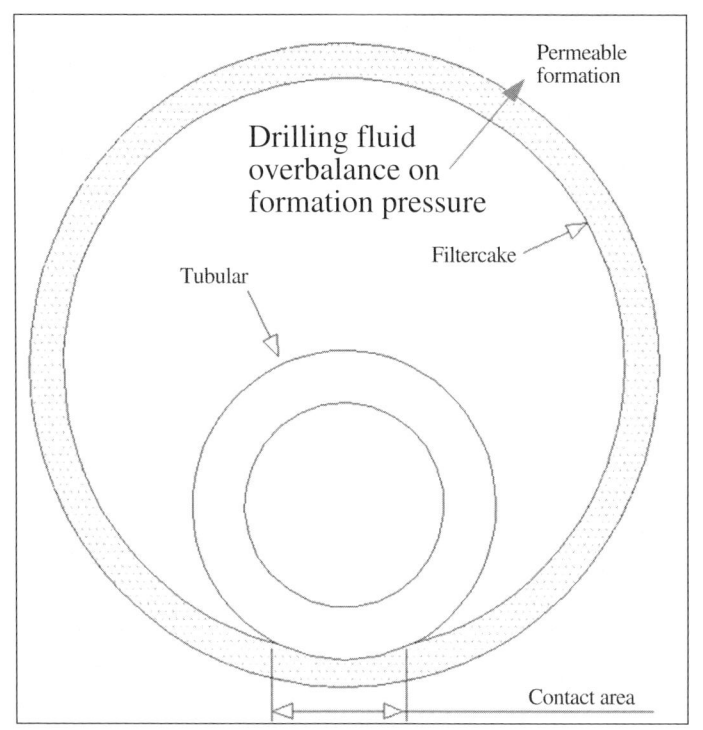

Fig. 13-2 Differentially Stuck Pipe

With wall contact over a period of time (seconds to minutes), the stationary tubular pushes into the filtercake. This establishes a contact area over which differential pressure may act. Larger diameter tubulars form greater contact areas and therefore are more likely to get stuck. Circulation will be possible but there can be no pipe movement until the pipe is freed.

The force required to pull the tubular free is proportional to both the differential pressure and the contact area. If this force exceeds the allowable pull on the pipe, then the differential pressure must be reduced or the filter cake thickness (and therefore the contact area) must be reduced. Casing is a particular problem because it is a large diameter, smooth pipe giving lots of good contact area with the formation.

To prevent differentially stuck pipe, the mud should be tailored so that it forms a thin, tough, non-sticky filter cake. Overbalance should be

Drilling Technology in Nontechnical Language

minimized, commensurate with maintaining primary well control. Stabilizers can be run to hold the large diameter drill collars off the wall. Drill collars can be used that have spiral grooves cut into them, which reduce the contact area. Finally, whenever the BHA is opposite a permeable formation, the time it spends static (neither reciprocating nor rotating) should be minimized (Fig. 13-3).

Sometimes, the drillstring or casing will get stuck by another mechanism elsewhere in the hole and then become differentially stuck if the conditions exist.

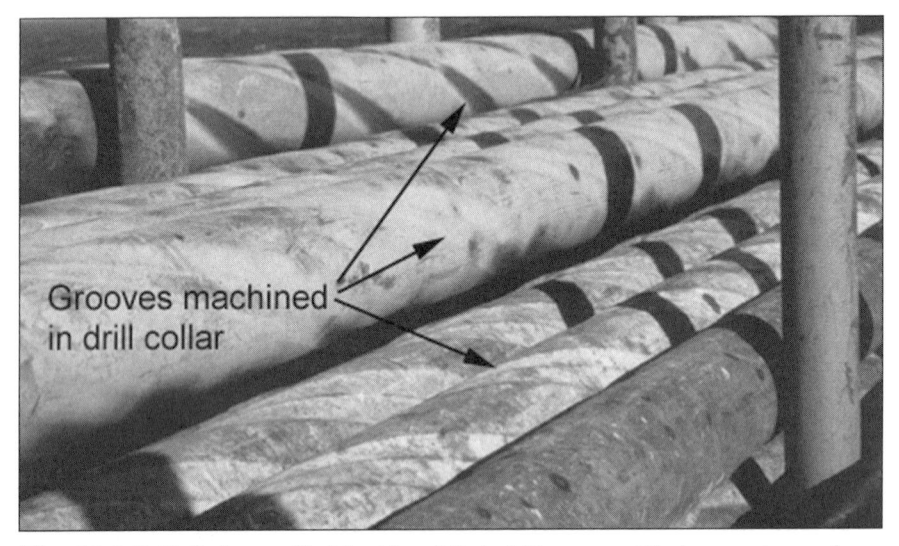

Fig. 13-3 Drill Collars with Machined Spiral Grooves. Photo courtesy of Schlumberger

Once pipe becomes differentially stuck, chemicals can be pumped down the drillstring and over the sticking formation to dehydrate and shrink the filter cake. Reducing overbalance is trickier; first, because of the well control implications and second, because significantly reducing the mud hydrostatic can have a highly destabilizing effect on some shales. The other action that can be taken is to pull and *jar* on the pipe; this is described next.

Jars and jarring

Steel, as a material, is quite elastic. Elasticity is a property of material whereby a force applied to the object causes it to deform and when the force

Chapter 13 • Drilling Problems and Solutions

is removed, the object reverts to its original dimensions. Springs are made of steel precisely because it is an elastic material. If more than a certain force is applied, the material stretches permanently—this force is called the *elastic limit*. If even more force is applied, the material will break—this force is called the *ultimate tensile strength*. Chapter 5 discussed how these limits effect the design of the drillstring.

When the drillstring is held in tension, it has stretched to a certain extent. How much it has stretched is measured by the *strain*. Strain is simply the amount that the pipe has stretched divided by the original length. So if a pipe of 10,000' long stretches by 5' then the strain will equal 0.0005. The energy that is used in creating this strain is stored in the pipe—this is called the strain energy.

A catapult works by using rubber strips to store strain energy. A stone is placed in the catapult pouch, pulled back (to store energy) and let go. The stored energy pulls the pouch and stone, accelerating it forwards. The initial acceleration is fast (because the pull on the pouch is greatest) and the acceleration reduces as the stone moves forward. When the rubber becomes slack, the stone stops accelerating and has reached its maximum speed.

A drillstring and jar are like a downhole catapult. The jar is a tubular tool that is placed high in the bottom hole assembly. It can open and close just like a bicycle pump. However, there is a mechanism inside the jar that stops the tool from opening until some condition is satisfied. The condition might be that a certain amount of pull must be applied to the jar, or that a period of time has to elapse after placing the jar in tension. Once the condition is satisfied, the jar opens by typically around 22" and will not open further (Fig. 13-4).

Above the jar one or two drill collars are placed. The rest of the BHA stays below the jar. If the pipe gets stuck (below the jar) then strain energy is applied to the drillstring by pulling upwards on it. The jar suddenly opens, which allows the strain energy stored in the drillstring to accelerate the heavy drill collars upwards. When the jar is completely open, the weight of the drill collars moves quickly upward, exerting a hammer blow on the jar and everything below it. In the catapult analogy, the drill collar is the stone that gets catapulted.

A jar is almost always run as part of a drilling BHA. Drilling jars are very robust so as to withstand many hours of rotating, vibration, high internal pressure, and changes in pressure and temperature. Fishing assemblies also incorporate jars.

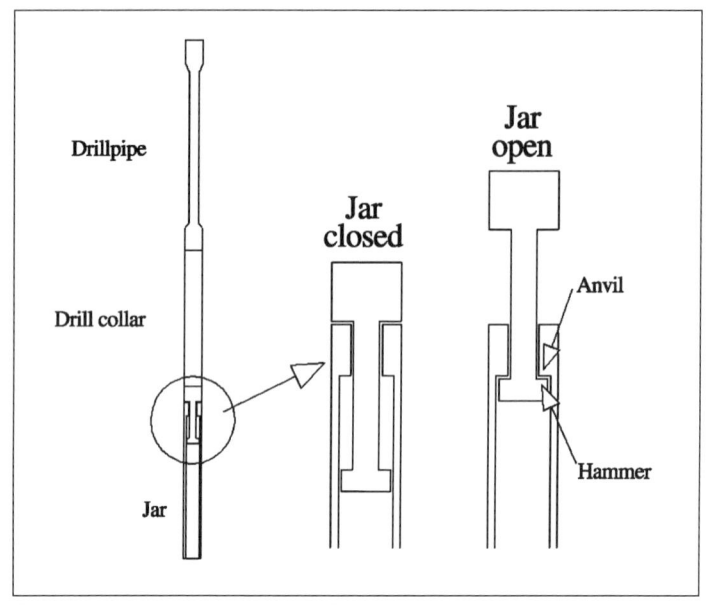

Drilling Technology in Nontechnical Language

Fig. 13-4 Jar—Principle of Operation

Fishing

Sometimes something gets into the hole that needs to be recovered. For instance, a spanner might drop down the rotary table or part of the drillstring might break. These items prevent normal operations from continuing. The item that must be removed from the well is called a fish and activities to remove the fish are termed *fishing*.

The following are four main causes for fish in the hole:

1. If there is a failure somewhere in the pipe in the hole that causes a break, the lower part of the string will drop into the hole.
2. If the stuck pipe cannot be freed, it has to be cut or unscrewed downhole.
3. Something falls into the well.
4. Sometimes wells are worked over—they may require that the existing completion is replaced with another completion. Often the completion tubing has to be cut downhole and recovered in pieces.

There are five classes of tools that can be used to remove a fish or junk from the hole. Which one to use will depend on the circumstances. The five classes are each briefly described.

Outside catch tools

If the junk has a round cross section at the top, the preferred tool will normally be an outside catch tool. These tools are pushed over the top of the fish. When the fishing tool is moved upwards it grips the fish and allows force to be applied.

There are two types of outside catch tools that are commonly available– the overshot and the die collar.

An *overshot* is very simple, strong, and versatile. It comprises a steel barrel—inside is a wedge shaped profile machined on the inside diameter. Fitting into this profile is an element that looks like a thick tube with slots machined in it, called a *grapple*. The grapple has a wedge-shaped outer profile that fits the ID of the outer barrel. On the grapple ID are hardened steel teeth that grip the fish. As pull is applied to the tool, the wedge shapes cause the grapple to close and so the more tension that is applied to the tool, the tighter the grapple grips the fish (Fig. 13-5).

Of all the fishing tools available, the overshot is overwhelmingly preferred to catch anything tubular shaped, because of the following:

- It can exert a tremendous pull on the fish. A jar can be placed above the overshot so that hammer blows can be applied to the fish through the overshot.
- It can be unlatched downhole if the fish cannot move. Latching and unlatching does not damage the top of the fish significantly.
- As it is hollow down the middle, logging tools and explosives can be run on wireline into the fish.
- If the fish is not plugged off, it is possible to circulate down the overshot and through the fish—a major advantage.

A *die collar* is a tubular piece of steel. Inside it has a slightly conical profile facing downwards. Threads are cut into this profile. To latch onto the fish, the die collar is lowered over the fish and a little weight applied. By rotating the drillstring to the right, these threads cut into the top of the

Drilling Technology in Nontechnical Language

fish and grip it. However, once the die collar is on the fish, it cannot be removed. It is not as strong as the overshot so if it were jarred on, it will likely pull off the fish.

Fig. 13-5 Overshot Grapple

Inside catch tools

Sometimes, a tubular fish has a large outside diameter, which prevents an overshot from going over it. Casing is a good example. For these fish, the fishing tool must grip on the inside diameter. There are two types of tools commonly available– the releasing spear and the taper tap.

The *releasing spear* works like an overshot in reverse. It has a mandrel down the middle, which has a conical profile outside. Steel slip elements fit around this mandrel with teeth on the outside. As force is applied, the mandrel profile pushes outwards on the slip elements, so as the more tension is applied, the more the slips grip. As with an overshot it can be released downhole.

The *taper tap* works like a die collar in reverse. It has a slightly wedge shaped section of bar below, which has threads cut onto the profile. The tap

Chapter 13 • Drilling Problems and Solutions

is lowered into the inside diameter, some weight is set down and the tap is rotated. The threads cut into the top ID of the fish and grip it. The taper tap is not very strong and as with the die collar, it cannot be unlatched once on the fish (Fig. 13-6).

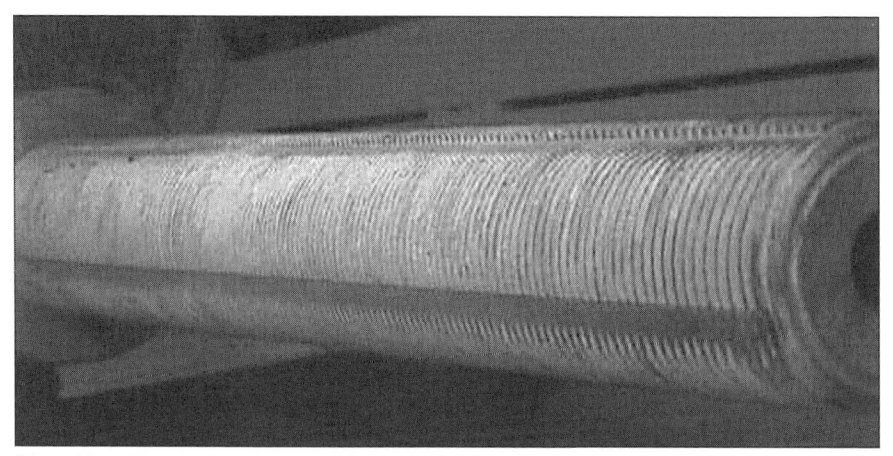

Fig. 13-6 Taper Tap

Washover and basket tools

Within this class of tools there are three types: the fishing magnet, tools that use fluid movement to catch small bits of junk, and tools that use a barrel to go over and closes underneath bits of junk.

A *fishing magnet* can be either a permanent magnet run on drillpipe or it can be an electro-magnet run on wireline. It is useful only for small pieces of ferrous junk.

Tools that use *fluid movement* include the *junk sub*. This tool is run just above the drill bit to catch small bits of

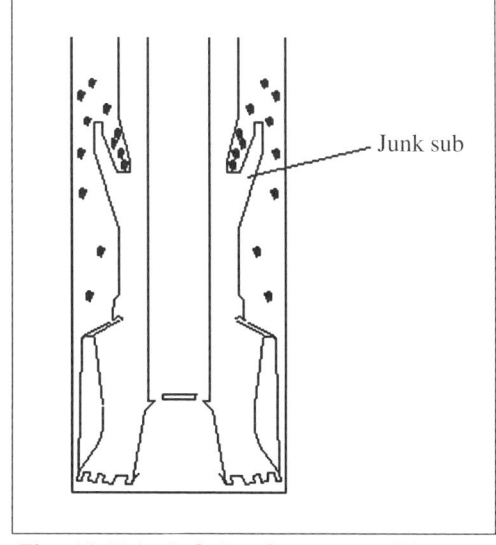

Fig. 13-7 Junk Sub—Conceptual Diagram

Drilling Technology in Nontechnical Language

junk that may have been left on bottom. It works because the flow of fluid up the annulus slows down slightly as it passes the lip of the sub, allowing heavy bits of junk to fall into the basket of the sub (Fig. 13-7).

Tools that use a barrel to go over the fish are generally called *washover* tools.

Washover tools are available with a milling shoe on the bottom (which can cut some rock at the bottom as well as mill on junk) and with spring loaded catch fingers inside. This is used to go over the junk and to catch it inside the barrel (Fig. 13-8).

Fig. 13-8 Washover Milling Tool with Catch Fingers

A simple washover tool can be made on the rig from a piece of small diameter casing or liner—see Figure 13-9. This tool is set on bottom, weight is applied and the string is rotated. This causes the fingers to bend and twist to close the fingers below the bits of junk. It is useful for catching rockbit cones or other small bits that are lying on the bottom of the hole (Fig. 13-9).

Chapter 13 • Drilling Problems and Solutions

Fig. 13-9 Homemade Washover Tool

Drilling, milling, and smashing junk

In the surface hole where formations are soft, smaller bits of loose junk can be drilled with a *steel tooth bit*. Often what happens is that the bits of junk get pushed into the side of the hole and are never seen again. Steel tooth bits are also capable of breaking up small bits of junk—no other drill bit can do this.

Mills can be used when junk is in the hole. Mills work well on junk that is not loose—is cemented in place or otherwise prevented from rolling around. Rolling junk can damage the mill (though mills can deal with this kind of junk also)—it might take more than one mill to do the job.

Mills for cutting junk (and especially junk that might be loose) use chips of tungsten carbide brazed to the bottom of a steel body as in Figure 13-10.

Explosives can also be run on wireline or pipe to break up junk, which can then be dealt with using a washover tool or a drill or mill. The explosive used is a shaped charge, designed so that the force is projected downwards onto the junk.

Fig. 13-10 Flat-bottomed Mill—for Milling Junk on Bottom

Fishing for wireline and logging tools

Finally, wireline sometimes breaks, resulting in a tricky fishing job. Wireline is fished with a *spear* that has barbs on the side. The spear is run into the top of the wire and rotated. The barbs catch on the wire and cause the wire to wrap around the spear so that it can be pulled out.

If a logging tool gets stuck, the best way to fish it is by *stripping over* the wire. The wire is cut at the surface and is fed through a special tool that latches onto a profile at the top of the logging tool. This fishing tool is then run in the hole on drillpipe over the wire so that the wire guides the fishing tool to the logging tool. It is a slow job because each stand of pipe has to have the wire fed through it before it can be connected to the string at the rotary table.

Economics of fishing

Fishing is an expensive activity because normal operations cannot continue until the situation is resolved. If the fish cannot be recovered, the alternatives are to:

- Place cement on top of the fish and drill a sidetrack around it

- Finish the well early (if enough of the original objectives can be achieved in the shallower wellbore)

- Abandon the well (either drill another well or abandon the project)

These alternative courses of action have a cost attached to them. There comes a point when fishing should be discontinued and the best alternative pursued. This is an economic decision.

Fishing jobs have the best chance of success on the first attempt. If the first attempt fails, it is only worth trying again if on the first attempt, something was learned that would improve the chances of success on the second attempt. The longer a fish stays in the hole, the harder it becomes to recover it. This is because hole conditions deteriorate with time, sticking forces will increase with time and as more attempts are made, the fish might become damaged or worn.

Fishing considerations for radioactive sources

If radioactive materials are contained in the fish, many governments mandate that they must be recovered even if the cost is very high. In this case, fishing can only be abandoned if certain rules are met (which will change with the area/government involved) after having exhausted all possibilities to recover the fish.

Some logging tools that are run on drillpipe contain radioactive sources and are designed so the source itself can be recovered. A special tool is run on wireline into the drillstring where it can latch onto the source container and pull it back to the surface.

Chapter Summary

This chapter looked at some of the commonly encountered downhole problems—lost circulation, stuck pipe, and fishing operations. When these problems occur, the root causes plus the current status of the well must be established before an effective strategy, specifically targeted to solve the problem, can be formulated.

Almost every drilling decision ultimately has an economic basis—which course of action will allow the well objectives to be achieved at the lowest overall cost. It might be tempting when problems occur to try to take short cuts to get back on track, but this is definitely not recommended. A professional engineered approach to solving the problem must be taken, not only to solve the problem safely and efficiently, but to prevent the situation from getting even worse.

Chapter 14

Safety and Environmental Issues

Chapter Overview

Management of safety is of prime importance to the oil industry. Over the years, systems have been developed to ensure a safe working environment for the people who work at the wellsite. In addition, many governments have become involved as a result of incidents that have occurred. In some cases, such as the Piper Alpha disaster in the UK North Sea, a single incident is so serious that public inquiries are held. This can lead to large changes in the legal requirements that govern the industry.

Ultimately, the safety of a person working at the wellsite is very much his or her responsibility—not to work unsafely nor allow others to do so and to alert supervisors to any unsafe conditions they see. However, the systems must be in place that lead to a safe environment and allow people to be effectively trained to work within those systems. The actions and attitudes of management must ensure that people are not encouraged—or intimidated—to work unsafely.

Safety Meetings

Meetings are an essential part of the safety regime on any rig. They are used to ensure good communication between everybody on the wellsite. There are several different types of safety meetings that are regularly held, as outlined below.

Pre-spud meetings

Once the drilling program is finalized and before the rig starts to drill, there will be a pre-spud meeting held in the office. All of the contractors (including the drilling contractor) will be represented at this meeting, which is generally chaired by the senior drilling engineer responsible for planning and supervising the operation from the office. The well design and drilling program will be discussed, and as part of this, any potential hazards will be flagged up. Potential risks must be managed, preferably by avoiding them altogether, but there must be procedures in place that will give a pre-planned strategy to minimize the impact and to recover from the effects of it as safely and efficiently as possible. Recognizing and managing risks is one of the primary responsibilities of everybody involved in the operation. The pre-spud meeting is probably the earliest forum where all concerned can discuss these risks.

On the rig, before the well is spudded, a similar pre-spud meeting will be held with the drill crews and rig supervisors, chaired by the drilling supervisor. This ensures that everybody is aware of the potential risks and how they will be addressed. The meeting is also a forum for feedback—a lot of practical experience will be represented in the room and good suggestions often result.

Weekly safety meeting

Every rig operated by a responsible company has weekly safety meetings that all personnel are expected to attend. These meetings generally take between 30 and 60 minutes. Items discussed will typically be the following:

- Any incidents or accidents that have taken place on the rig
- Any incidents or accidents on other rigs that may be particularly relevant to the rig
- Suggestions from the participants to improve safety on the rig
- Status of any ongoing work that has a bearing on safety
- Any STOP cards that have been handed in can be discussed and the Lessons Learned emphasized (STOP cards are discussed later in this chapter)
- Rig safety statistics

Chapter 14 • Safety and Environmental Issues

These meetings are recorded and the minutes are sent to the office, as well as being posted on the rig notice board.

Daily operations meeting

On some rigs, a daily meeting is held in the morning to discuss work for the day. Generally, the attendees might include the toolpusher, drilling supervisor, driller, mechanic, and electrician. While enjoying a cup of fresh coffee and donuts, the anticipated program over the next 24 hours will be looked at. Any planned maintenance or repair work will be coordinated so that normal rig operations are affected as little as possible; and the risk of something conflicting is reduced.

Pre-tour meeting

Before starting their 12 hour shift, the driller will call together the drill and deck crews. The work program for the shift will be discussed and any special considerations noted. This allows the new shift to start work knowing what the overall plan is, what hazardous operations might be coming up, and what equipment needs to be prepared.

Pre-job safety meeting

Before starting any non-routine work, the onshift driller will call a safety meeting of all the people who will be involved in the work. As the driller has overall responsibility for the safe execution of drill floor operations, it's important that he runs the meeting.

Here is an actual checklist for a pre-job safety meeting for running casing. Present at this meeting would be the drill and deck crews and any contractors involved in running the casing. The toolpusher and drilling supervisor might also be present.

Pre-job safety meeting for running casing

1. A good safe job is required; work efficiently without rushing.
2. All the correct safety equipment is to be used: safety belts at height, etc.–there will be no exceptions.
3. The on shift driller has overall responsibility for the job. Any problems are to be reported to him.

Drilling Technology in Nontechnical Language

4. Watch out for pinch points: fingers or hands getting trapped between moving and stationary equipment, such as casing coming up to the drill floor.
5. Keep the drill floor reasonably clean to prevent tripping and slipping hazards.
6. No one is to use the V-door stairs when a joint of pipe is being moved between the catwalk and the drill floor or being picked up by the single joint elevators.
7. When you change shifts, make sure you hand over your job to your relief, then stop and watch them working for a few minutes to make sure they are doing the job properly.
8. Any unsafe conditions must be corrected immediately. Watch out for ropes on the V-door and stabbing board getting worn and change in good time. Keep an eye on lifting slings and straps.
9. Joints of casing must have clamp-on protectors in place on the pin end before picking up the V-door. If one drops off, the rig crew must be told straight away and the pin examined before running.
10. It must be possible to circulate the casing in case of a well control situation or if the casing has to be washed past a tight spot. The crossover from casing to the high pressure pump line must be kept ready to use on the drill floor.
11. The men working on the stabbing board must be rotated out every couple of hours and they must also get adequate rest[1].
12. Highlight any special procedures, hazards, etc., with the crews.
13. Invite questions or comments from the crew.

Newcomers on the Rig

When people arrive at an offshore rig, they usually sit through a briefing about the rig. They will probably see a video about the rig showing the rig layout, lifeboat locations, etc. The video will demonstrate the various alarm sounds, what they mean, and what actions are to be taken on hearing them. They will also be told when the weekly safety meeting will be held and that they must attend if they are on the rig at that time.

Chapter 14 • Safety and Environmental Issues

They will be assigned a cabin and a bed (even if they are only day tripping) and told which lifeboat to use in an emergency.

If they are new to the rig, they will sometimes be given a distinctive green hard hat. This marks them out so that everybody else knows that they are not familiar with the rig.

Training and Certification

Safety training is an essential part of any safety regime. Courses will be organized by the drilling contractor and by the operator for rig personnel to attend during their time off the rig. Some of these courses have to be repeated at regular intervals, such as the Well Control certificates mentioned in Chapter 11.

Survival and firefighting is covered usually in a 5 day course and teaches how to escape from a submerged helicopter, deploy and enter liferafts and lifeboats, and how to fight various types of fires. This is a practical course; attendees are strapped in to a mock helicopter cabin that is ditched in to a pool and turned upside down. Divers are in the water to assist anybody who has real problems. In the mock cabin, all the windows are out and the door is open, but even for somebody who's at home under the water it's a little stressful. It's easy to see that in a real ditching and especially in rough water, if the chopper turned over and sank straight away you'd have a real problem getting out. One thing that would aid escape would be to carry a mask and snorkel—as any diver will know, human eyes do not focus when submerged in water, but wearing a mask allows proper vision (though everything will appear bigger because of the refractive index of water). It would also be worth carrying a diving knife that has a thick, strong blade and can be used as a lever, however attitudes in some places will make it hard or impossible to carry such a "weapon" in the cabin. Carrying something that can significantly increase your chances of survival in an emergency situation should not be discouraged.

In most developed areas it is also necessary to have a medical certificate. These certificates are valid for five years up to age 40, and then for three year periods after that.

If Hydrogen Sulphide (H_2S) might be present in the well, a one or two day course will be organized. This will teach or remind attendees about the characteristics of the gas, what to do in an H_2S emergency, how to use the self contained breathing apparatus (SCBA) and escape sets, as well as how to work in an H_2S environment (as opposed to escaping through it).

Drilling Technology in Nontechnical Language

If other characteristics of the well demand it, customized training courses may be arranged. An example would be drilling a high pressure high temperature well where the rig to be used and the crews on it have little or no experience of that kind of well.

Drills

It is all very well to attend courses once in a couple of years, but any skill gets rusty if not practiced regularly. Drills give personnel the chance to practice what to do in various situations. Drills are held regularly on the rig to simulate a variety of potential problems.

Fire drills

The fire alarm is sounded and the designated fire teams are sent to an area of the rig. There, they will deploy fire hoses and perhaps rescue one or more "victims". Meanwhile anybody not on duty on the drill floor or in other areas will report to a defined muster point on the rig, where they will be marked off against the POB list. The medic will prepare the rig hospital for casualties.

Abandon rig drill

May be done as part of the fire drill. The Toolpusher or OIM will change the fire alarm to the Abandon Rig alarm and everybody at the muster point will get into the lifeboats.

Kick drill (also called a pit drill)

The toolpusher or drilling supervisor will cause an indication to the driller that the active pit level is increasing (there are several ways to achieve this). The driller is expected to stop drilling and make a flow check. Once a flow check is under way he will be told that this is a drill and he is to simulate closing the BOP. The reaction time of the driller and crew will be noted.

Trip drill

While tripping in or out of the hole, the toolpusher or drilling supervisor will tell the driller to assume that the well is flowing. The driller will lower the drillstring so that the crew can place a valve on top of the drillpipe. This valve is closed, then the BOP is closed and the well secured. The total time from initiating the drill to securing the well will be noted.

A drill crew can expect either a trip drill or a kick drill each week, without any prior notice.

H$_2$S drill

If working on a well where H$_2$S might be present, H$_2$S drills will be held regularly. The alarm will sound, personnel will either don a SCBA set (if working) or will take an escape set and report to the upwind muster point.

Permit to Work Systems

A permit to work (PTW) system ensures that non-routine work is carried out safely and without conflicting with other work going on around the rig. A permit is signed by the OIM or toolpusher–sometimes the drilling supervisor will countersign the permit.

The permit will contain the conditions under which the work must take place, any safety precautions required, how long the permit is valid, and any other relevant information. Once the work is complete, the permit is returned to the OIM and closed.

Permits are generally issued for the following work:

Hot work. Any welding, cutting, or grinding that will be done in an area that could potentially experience an explosive atmosphere. Typically this will be anywhere outside of the welder's workshop or accommodation. The hot work permit will call for gas detection tests to be done before the job and perhaps at intervals during the job and any special firefighting precautions (such as keeping an extinguisher or hose ready at the job site).

Tank entry. If somebody has to enter a tank or enclosed space, the permit will call for the atmosphere to be tested to ensure it is breathable. It is likely that the person entering an enclosed space must wear breathing apparatus. A safety monitor will stay outside the tank, probably with a radio transmitter and if the person in the tank requires assistance, this monitor will call for help first and then take appropriate action to help.

Over water work. If somebody has to work over the water except on properly constructed walkways, several precautions will be taken. A safety harness with a line attached to the rig structure will be required, as will a flotation vest. A safety monitor will stand by with a radio. The standby boat will be called and will stay down current of the rig for as long as the work continues.

Electrical work. Where electrical equipment will be worked on outside of the electricians workshop, a work permit is needed. This permit will show how the equipment is to be electrically isolated from the supply, plus any other precautions that may be required.

Safety Alerts

Various agencies (governments and companies) issue notices to the industry to publicize incidents or accidents. The purpose of disseminating safety alerts is that lessons can be learned that may prevent a future occurrence of a similar incident.

While government agencies disseminate safety alerts freely, companies are rather more reluctant to place this information in the public domain. This is because if it identifies an incident that occurred on a company rig or operation, it might lead to bad publicity. It would be preferable if, instead of merely disseminating this information within the company, the responsible government authority could issue the safety alert, but disguise the source of the incident. The objective of disseminating the alert can be achieved without identifying the rig and well on which the incident occurred (Fig. 14–1).

Equipment Certification

Certification by a responsible and recognized authority assures users that a particular piece of equipment is fit for purpose. Of special interest are items involved in lifting (such as slings, shackles, wire ropes, and lifting frames) and pressure vessels (such as gas cylinders, pressurized storage tanks, and high pressure lines) because if these fail, the likelihood of serious injuries or fatalities is quite high.

Lifting equipment is generally tested every three or six months. Slings on a rig are often color coded with paint–if a sling does not have the current color painted on it, it should not be used.

Pressure vessels bear a metal plate that an inspector will stamp after visually inspecting the tank and applying test pressure to it. Many people do not appreciate the effect of a pressure vessel bursting with low pressure

Chapter 14 • Safety and Environmental Issues

Example of a Safety Alert

Shallow Gas Blowout

Issued by: Minerals Management Service (MMS), USA
Date of Issue: January 5, 1984
Issuer's serial number: Notice No. 123

A Conductor hole was drilled to 975 feet on a platform well in 313 feet of water and the drill pipe was being tripped out of the hole. After pulling five stands of pipe, without filling the hole, the mud pump was started to fill the hole. While pulling the sixth stand with the mud pump running and only 11 strokes on the counter, the well started flowing (bit at 439 feet MD). The annular BOP was closed, one of two diverter lines were opened, the drill pipe safety valve was installed, and all engines were shut down.

The crew abandoned the platform by crossing a connection bridge to another platform. Gas, salt water, and sand were observed blowing out of the diverter line. The well bridged over and ceased blowing after approximately 45 minutes. The crew returned to the platform and turned on one generator for lighting, and the rig and platform were washed down. There was no sign of pollution nor any fires or personnel injuries. The diversion was considered a success.

In order to safely drill the conductor hole section of future wells planned in this area, the operator recommends the following precautionary procedures be strictly observed:

1. The Drilling Foreman (or Supervisor) should be on the drill floor at all times while the drill pipe is being pulled.

2. Prior to pulling the drill pipe, check for evidence of underbalanced conditions or balled-up bit and circulate until the drill pipe will pull without swabbing.

3. Pull the drill pipe slowly to prevent swabbing. Stop pulling pipe and completely fill the hole at frequent intervals to minimize reduction in hydrostatic head and assure that the swabbing is not occurring.

[signed] D.W. Solanas
Regional Supervisor
Rules and Production

2-164

Fig. 14-1 Example of a Safety Alert

on it—only a few psi is enough to exert a force on tank plates that will cause large pieces to fly off with enough force to kill.

Electrical equipment carries a rating that signifies whether or not it can be used in potentially explosive atmospheres. A drilling rig (or production station) is divided into zones according to the likelihood of an explosive atmosphere being present. A "zone 1" area is likely to experience an explosive atmosphere in normal daily operations; a "zone 2" area may experience an explosive atmosphere if abnormal events take place, and a "zone 3" area is unlikely to be exposed to an explosive atmosphere. Outside of the accommodation, most areas of the rig will be rated "zone 2" and so the electrical equipment used (including switches, motors, and power tools) must be appropriately certified.

Safety Equipment

Safety equipment can be divided into two groups—*Personal Protective Equipment* (PPE) and other safety equipment.

Personal protective equipment

On most rigs, everybody working outside of the accommodation or workshops must wear safety boots (with steel toes), hard hats, and safety glasses. Some companies also demand that approved, flame resistant overalls be worn.

For specialist tasks such as welders, floormen and derrickmen, other items of PPE will be issued that are related to the task in hand. Such items would include welding gloves, and goggles, work gloves, and safety harnesses.

For personnel who must be exposed to chemicals while mixing mud or cement, PPE may include rubber gloves, rubber safety boots, rubber aprons, goggles, and breathing masks.

Other safety equipment

Also available at various places around the rig will be equipment that is available to be used if a problem occurs. In areas where toxic or corrosive chemicals are stored or mixed, eyewash stations must be placed where they can be easily reached. These must be regularly checked to ensure that the water container is full, rubbish is not piled up on top or around the station, and that it's in good condition overall. Showers might also be placed in chemical storage areas.

Fire extinguishers, hoses, and axes will be placed at strategic positions.

Safety notices and barriers alert rig personnel to particular hazards, such as high pressure lines.

Temporary barriers are erected to keep non-essential personnel away from temporary hazards, such as high pressure testing.

STOP

The DuPont company some years ago designed a program to reduce accidents and incidents in it's factories. This program is called *STOP* and it has been implemented on many rigs.

Chapter 14 • Safety and Environmental Issues

The idea behind *STOP* is that if somebody witnesses an unsafe act (some action that may lead to an accident), they should do the following:

1. Inform the person doing the unsafe act that they should *STOP* doing it. This must be done in a non-confrontational manner.

2. Fill in a card that describes the unsafe act, but does not name the person doing it. The person filling in the card can optionally put his or her name on the card, but is not obliged to do so.

3. Somebody will be responsible for collating the *STOP* cards and seeing what lessons can be learned.

4. The lessons learned are disseminated at the weekly safety meetings. If a number of *STOP* cards show a recurrent theme then the safety manager may decide that the problem may be widespread and some extra action is warranted—such as arranging training courses or writing articles in newsletters.

STOP has proved to be very effective in reducing serious incidents. It's use is quite widespread in the industry.

Minimizing Discharge and Spills

The oil industry gets a lot of bad press about discharges of substances in to the environment. While this type of criticism would have been fair enough ten years ago, it is very different in most places at the end of the twentieth century. Discharges from rigs owned by responsible companies are tightly controlled. Modern rigs have systems designed and built in to the rig to minimize any discharge from the rig that does not meet strict standards of cleanliness.

In areas where no special environmental considerations apply, some discharges can be made without adversely affecting the environment. In sensitive areas, solid and liquid wastes can be disposed of in other ways.

Offshore rigs

Solid waste from rigs includes drilled cuttings. Where water-based muds are used, these cuttings can often be safely discharged overboard where they simply pile up on the seabed. In sensitive areas where even clean cuttings (that is, without any hydrocarbons present on them) might damage the environment, cuttings can be disposed of in other ways, as can cuttings with oil on them. First, they can be collected in containers and shipped to land for disposal in a landfill. Second, they can be ground up into a slurry and injected into one of the casing annuli, if suitable formations are exposed. Food waste is generally ground up so that it will pass through a coarse mesh and dumped overboard, but again in sensitive areas this can be shipped to land.

Sewage can be treated so that it can be safely discharged overboard without pollution.

Liquid discharges can be somewhat trickier due to the volumes involved. Water-based muds and cements can usually be discharged overboard as no lasting harm will result. Oil-based muds based on mineral oils are not discharged, but are stored and often recycled, resulting in savings to the operator.

Scrap steel, rope, plastic containers, medical waste, and other rubbish are not discharged, but are returned to the shore base for responsible disposal.

Onshore rigs

As part of location preparation, land rig locations have waste pits dug. One of these waste pits—the largest—is positioned next to the mud tanks. Drilled cuttings and mud are dumped into this pit. It's a good idea to dig a large U-shaped pit and to dump mud and cuttings at one end. At the other end of the U, the liquid that drains off can be sucked out—sometimes it can be recycled into the mud system (as it is full of expensive chemicals) and this is pretty useful in arid regions where water supply is expensive. Alternatively it can be sucked in to a tanker and disposed of or processed elsewhere.

After the well has finished, the pit is allowed to dry out. This can take several weeks, even in the desert. Once it has dried, the pit is re-covered with the originally extracted earth and the site is restored. In many cases the wellhead can be visited and it is difficult to tell where the waste pits were.

Sewage pits are also dug and pipes are run from the accommodations to this pit. As with the rig waste pit, this is allowed to dry out before being covered over and the site restored.

Inadvertent spills

Sometimes an accident or incident will occur that places large volumes of polluting fluids into the environment. This may be crude oil, rig fuel, oil based mud, or some other liquid pollutant. This will initiate a pre-planned spill response that may involve other operators in the area, the coast guard, and other public agencies and various contractors.

In many areas, the operators active in the area get together and share the cost of a spill response team. Equipment is stored in areas where response vessels can quickly load up and head out to the area. This equipment will include booms (to contain the spill), pumps, tanks, and chemicals.

To have all this equipment standing by costs quite a lot of money; it is hoped the equipment will never be needed, but nevertheless is part of the operators' responsibility.

Environmental Impact Studies

Even in areas that are not especially sensitive, many operators carry out studies to gauge the impact of their operations on the flora, fauna, and aquatic life around the operation. This assessment will cover not only the result of normal operations, but also what would happen if an oil spill occurred. The effects of prevailing winds and currents will be studied to see which areas would be at greatest risk of a landfall so that contingency plans can be drafted.

In sensitive areas, public inquiries might be held so that the public can understand the operator's plans and can comment and raise concerns.

Sometimes there are alternative courses of action that can reduce the environmental impact of an operation. BP drilled record breaking horizontal wells to exploit a reservoir offshore by drilling from land on the south coast of England. These wells had thousands of feet of horizontal wellbore successfully drilled from a land rig, which avoided using an offshore rig or creating artificial islands (which was one of the alternatives studied when the field was discovered). Much attention was paid to the environmental impact of the various alternatives and the result was a successful exploitation of the field without upsetting anybody!

Severe Weather—Suspension of Operations

Severe weather can create significant danger for rig crews. In areas prone to hurricanes and tornadoes, such as the Gulf of Mexico at certain times of the year, close attention is paid to the possible courses that such storms might take. When it becomes possible for a rig to be hit by one of these storms, the well is secured and the rig abandoned.

In other areas other problems might prevail—such as sandstorms. The judgement of the drilling supervisor must be used to ensure that operations are suspended before weather conditions get so dangerous that the likelihood of an incident increases significantly.

Chapter Summary

Safety management is as important as managing any other aspect of the operation. Every responsible manager and supervisor in this industry wishes to see everybody return home safely to their families at the end of their time on the rig. Most companies have very clear policies and statements that ensure that everybody concerned knows that unsafe practices will not be tolerated.

This chapter set out many of the safety concerns that result from operating drilling rigs. Equipment, training, practices, and procedures are used to reduce risk to as low as reasonably practical (ALARP).

There is no longer room in this industry for anybody who does not take safety and environmental protection seriously.

Safety is *NO ACCIDENT*.

Glossary

[1] **Note from the author.** I encountered an incident once where a member of the team of contractors who were on the rig to help run a long string of casing actually fell asleep while working on a small platform about 35' above the rig floor. Investigation of the incident showed he had been working for around 30 hours with only short breaks for meals. It was clearly his responsibility to himself and his colleagues to insist on taking a proper break for sleep before he reached such a state. It was also the responsibility of the drillers and of his contractor supervisor to insure this. He was not in danger of falling because he wore the required safety

Chapter 14 • Safety and Environmental Issues

harness and line; however it's easy to see that he could have presented a danger to people working around (and below) him.

[2] **POB.** Persons on Board. A list that is kept up-to-date and sent into the office daily; shows the name, company, assigned room, and lifeboat of everybody on board.

[3] **SCBA set.** Self Contained Breathing Apparatus. Comprises a compressed air bottle on a harness, a full face mask, and a demand valve.

INDEX

A

Abandon rig drill, 312
Abandonment of well, 73–74, 312
 removing test string, 73–74
 well safety, 74
 definition, 74
 safety drill, 312
Accidents (discharge/spill), 319
Accommodation, 106–107
Accumulation (oil and gas), 22
Acid test, 218
Active mud system, 116–118
Active system, 261
Aerated/foamed mud, 151
Air as circulating medium, 150
Anderdrift vertical inclination indicator, 186
Andergauge steering tool, 180
Angular unconformity, 20
Annular velocity, 50–52
Anticline, 20
API gravity, 235
Appraisal, 100
Assistant driller, 267

Axial–internal forces, 196

B

Bag type blowout preventer, 239–241
Baker Hughes Autotrak, 180
Barite, 131
Barrels, 76
Bedding plane, 7
Bingham plastic fluids, 157–158
Bit economics, 143–144
Bit grading, 141–142
Bit selection, 142–143
Bit sub, 52–53
Blind ram blowout preventer, 243–244
Block (area), 27–28
Block line, 48
Blowout (well), 13
Blowout preventer, 9, 46–49, 57–60, 114, 239–247
 stack, 239–247
Blowout preventer stack, 239–247
 bag type (annular), 239–241
 ram type, 241–244
 choke valves, 244–245

control systems, 245–246
subsea systems, 246–247
Bottomhole assembly, 53–54, 62
Bottomhole coring, 220–224
  sleeve coring, 221
  sponge coring, 221
  orientated coring, 221–222
  pressure coring, 222
Bottoms up, 101
Boyle's law, 262
Brine, 77
Brine mud, 149–150
Buckling, 195–196
Build assembly, 179
Bullnose, 56
Burst strength/stress, 58, 194

C

Calcium carbonate, 5
Caliper logging, 96–97
Camp boss, 267
Capillary forces, 26
Carbon dioxide, 25
Carbonates, 5–6
Casing (well), 34–36, 42, 56–58,
    64–68, 82–85, 189–214
  string design, 34–36, 42, 192–197
  running/cementing, 56–58, 64–68,
    189–214
  size, 82–85
  functions, 190–192
Casing cementing, 189–214
  cement functions, 198

mud removal, 198–199
cement, 200–201
cement design, 201–208
casing running/cementing,
    208–209
surface casing, 210–211
cement evaluation, 211
secondary cementing, 211–212
lost circulation curing, 212
cement plugs, 212–213
Casing functions, 190–192
  conductor pipe, 190
  surface casing, 190
  intermediate casing, 191
  production casing, 191–192
Casing hanger, 64
Casing running/cementing, 56–58,
    64–68, 208–209
Casing shoe, 34
Casing size, 33–34, 82–85
Casing string design, 34–36, 42, 82–85,
    192–197
  tension, 192–193
  compression, 193–194
  burst, 194
  collapse, 195
  driving forces, 195
  buckling, 195–196
  temperature, 196
  axial–internal forces, 196
  corrosion, 197
  connections, 197
Casinghead housing, 64–65
Catch tools, 299–301
Cement, 35, 86–87, 198, 200–208,
    211–213, 218, 289–290
  design, 201–208

Index

Cement curing (lost circulation), 289–290
Cement design, 201–208
    density, 201–203
    thickening time, 203
    compressive strength, 203
    temperature rating, 203–204
    rheology, 204–205
    chemical additives, 205
    massive salt formations, 205–208
Cement evaluation, 211
Cement functions, 198
Cement plugs, 212–213
Cementation, 218
Cementing, 35, 86
Cementing conductor, 86–87
Chemical additives, 205
Chemical properties (mud), 164–167
    reactive shales, 165
    salts, 165
    reservoir damage, 166
    corrosion downhole, 166
    hydrogen sulfide problems, 166–167
Chert, 6
Chip hold down, 140
Choke valves, 244–245
Christmas tree, 72–73
Circulate the well, 76
Circulating pressure losses, 168
Clay minerals, 4–5
Cleavage, 217
Coal, 17
Coiled tubing, 103, 129, 231–232
    logging, 231–232

Collapse strength/stress, 195
Color, 217
Completion design, 30
Compression, 193–194
Compressive strength/stress, 8–9, 203
Conductor driving, 45
Conductor pipe, 38, 45, 49, 190
Coning, 81–82
Connections, 197
Consolidation (rock), 17
Contingency costs, 275
Continuous phase, 167
Contract types, 268–270
Contractors, 266
Control systems, 245–246
Core bits, 137–138
Core sample, 75, 138–139
Coring, 43, 220–224
Corrosion, 166, 197
Crane driver, 267
Crownblock, 121–122
Crystal shape/form, 217

D

Daily operations meeting, 309
Data gathering, 30–33
Decision making, 272–273
Deep formation lost circulation, 287–288
Density gradient, 262
Density (cement), 201–203

hydrostatic pressure, 201
  cost, 201
Density (mud), 207
Depth dependent costs, 275
Derrickman, 267
Design considerations (mud rheology), 162–163
Designing the well, 33–36
  hole sizes, 33–34
  casing design, 34–36
Development well drilling (offshore), 79–101
  well planning, 79–82
  hole/casing sizes, 82–85
  well program, 85
  drilling the well, 86–100
Diagenesis, 2
Diesel oil bentonite plug, 288–289
Differential sticking, 168, 294–296
Dilatent fluids, 160
Directional characteristics, 30–32
Directional profile, 84–85
Directional/horizontal drilling, 80, 171–188
  economic justification, 171–174
  single surface location, 171–172
  inaccessible location, 172
  salt dome drilling, 173
  multiple wells from borehole, 173
  onshore to offshore reservoir, 173
  horizontal well in reservoir, 173
  remedial work/sidetrack, 173
  relief well, 174
  tools/techniques (kickoff), 174–177
  wellpath control, 178–180

horizontal wells, 181–182
multilateral wells, 183–184
surveying, 184–187
navigation, 187–188
Discharge/spill minimizing, 317–319
  offshore rigs, 317–318
  onshore rigs, 318
  accidents, 319
Dispersed mud, 147–148
Displacement, 100
Diverter, 38, 46–49, 57–59, 75
  setup, 46–49, 57–59
Downhole motor with bent sub, 176
Downhole sampling, 220–225
  bottomhole coring, 220–224
  explosive sidewall coring, 222–223
  rotary sidewall coring, 223–224
  pore fluid sampling/analysis, 224–225
Downhole steerable motor, 176–177
Drawdown pressure, 188
Drill bits, 49–50, 133–144
  roller cone bits, 134–135
  fixed cutter bits, 135–138
  core bits, 137–138
  optimizing drilling parameters, 138–141
  grading the dull bit, 141–142
  bit selection, 142–143
  bit economics, 143–144
Drill collar, 49–50
Drillability, 101
Driller, 76, 267
Drilling barge, 109
Drilling contractor, 266

Index

Drilling derrick, 47–48
Drilling engineer, 264
Drilling fluids, 42, 145–169
 functions of, 145–146
 mud classifications, 146–151
 design of, 151–167
Drilling hydraulics, 141
Drilling manager, 264
Drilling notes, 41–42
Drilling out casing, 58–61
Drilling platform, 111–113
Drilling practices, 41
Drilling rig classifications, 104–113
 heavy land rig, 104–105
 light land rig, 105
 helicopter transportable land rig, 106
 semi–submersible, 106–107
 drillship, 107
 drilling tender, 108–109
 drilling barge, 109
 jackup rig, 110–111
 platform, 111–113
 submersible, 113
Drilling rig systems/equipment, 113–129
 dynamic positioning, 113–114
 high pressure pumping equipment, 114–116
 active mud system, 116–118
 solids–control equipment, 118–121
 hoisting equipment, 121–123
 rotary equipment, 122–125
 drillpipe, 125–126
 drillpipe handling equipment, 127–129

Drilling supervisor, 261, 264–265
Drilling surface casing, 87–89
Drilling tangent section, 94–95
Drilling tender, 108–109
Drilling the well (land), 44–70
 location preparation, 45
 conductor driving, 45
 ordering equipment, 45–46
 checking infrastructure, 46
 moving rig, 46–49
 diverter setup, 46–49, 57–59
 spudding the well, 48–52
 first hole section, 52–56
 running/cementing surface casing, 56–58
 blowout preventer, 57–59
 drilling out casing, 58–61
 intermediate hole section, 61–63
 logging, 63–64
 casing running/cementing, 64–68
 production hole section, 68–69
 logging, 70
 production liner running/cementing, 70
Drilling the well (offshore), 86–100
 spudding, 86–87
 cementing conductor, 86–87
 drilling for surface casing, 87–89
 heave compensator, 89–91
 kickoff/kicking off, 90–93
 drilling tangent section, 94–95
 locating casing point, 95–96
 logging, 96–97
 bringing to horizontal, 97–98
 horizontal drilling, 99–100
 suspending the well, 100
Drillpipe, 125–129, 231–232

327

handling equipment, 127–129
   logging, 231–232
Drillship, 107
Drillstring, 76
Driving forces, 195
Drop assembly, 178–179
Dynamic kill (well), 254–256
Dynamic positioning, 113–114, 130\

E

Economics (fishing), 305
Electrical potential tools, 230
Electrical logging, 225–232
   resistivity/induction tools, 226
   resistivity logs, 226
   induction logs, 226
   microresistivity tools, 226
   sonic tools, 226, 229
   radioactivity tools, 229
   mechanical tools, 229–230
   electrical potential tools, 230
   temperature log, 230
   wireline logging, 230–231
   drillpipe/coiled tubing logging, 231–232
   logging while drilling, 232
Electrical work permit, 314
Environmental impact, 32–33, 270, 307–321
   safety, 307–321
   studies, 319
Equipment certification, 314–315
Evaluation techniques, 215–236

physical sampling (surface), 216–220
physical sampling (downhole), 220–225
electrical logging, 225–232
production testing, 232–235
Evaporites (salts), 6–8
Event sequencing, 42
Exploration well drilling (land), 27–77
   prospect identification, 27–28
   well proposal, 28–30
   gathering data, 30–33
   designing the well, 33–36
   writing the well program, 36–44
   drilling the well, 44–70
   production testing, 70–73
   abandoning the well, 73–74
Explosive sidewall coring, 222–223

F

Filtrate, 168
Fire drill, 312
First hole section, 52–56
Fish, 12
Fishing, 298–306
   outside catch tools, 299–300
   inside catch tools, 300–301
   washover/basket tools, 301–303
   junk removal, 303–304
   wireline/logging tools, 304–305
   economics, 305
   radioactive sources, 305–306
Fishtail bit, 138
Fixed costs, 274

Index

Fixed cutter bits, 133, 135–138
Fixed pipe blowout preventer, 241–242
Flat time operations, 270
Floating production and storage offshore, 172, 188
Floating rigs (shallow gas), 256
Flow regimes, 161–162
Fluid loss (mud), 153, 168
Fluorescence, 219
Fractured/cavernous formations, 288
Fractures, 219
Free fall, 214

G

Gas drive, 23–24
Gas–oil contact, 75
Gelation, 167
Generation of oil/gas, 15–26
    source rock, 16–17
    rock properties, 17–19
Geological information, 43
Geology, 1–13
    rock types, 1–2
    plate tectonics, 3
    lithology, 3–8
    rock strengths/stresses, 8–9
    hydrostatic pressure, 9–12
Geometry sticking, 291–292
Gilsonite, 214
Grading the dull bit, 141–142

H

Hardness, 217–218
Hazard potential, 40
Heave compensator, 89–91
Helicopter transportable land rig, 106
Herschel–Buckley fluids, 160–161
High angle/horizontal well killing, 256–257
High pressure pumping equipment, 114–116
High pressure/high temperature wells, 257–259
Hoisting equipment, 121–123
Hold/tangent/locked assembly, 178
Hole cleaning ability, 169
Hole deviation, 30–31
Hole opener, 56
Hole sections, 52–56, 61–63, 68–69
Hole sizes, 33–34, 82–85
    directional profile, 84–85
Horizontal drilling, 97–100, 171–188
Horizontal wells, 173, 181–182
Hot work permit, 313
Hydrated shales, 4–5
Hydraulic horsepower, 115
Hydrocarbon generation, 16–17
Hydrocyclone, 119–121
Hydrogen sulfide problems, 24–25, 166–167, 313
    safety drill, 313
Hydrostatic pressure, 9–12, 35, 201

I

Igneous rock, 1–2
In gauge, 77
Inaccessible location, 172
Incentive schemes, 270–272
    safety, 270
    flat time operations, 270
    metering specific targets, 271
    percentage of well cost saved, 271
    attributes, 271–272
Incident response, 279–280
Inclination (well), 97
Induction logs, 226
Information on well program, 38
Infrastructure check, 46
Interfacing (service companies), 273–274
Intermediate casing, 57–58, 61–68, 191
Intermediate hole section, 61–63
Internal reservoir characteristics, 75
Invert oil emulsion mud, 150

J

Jackup drilling rig, 110–111
Jackup production platform, 172
Jars/jarring, 296–298
Jetting, 174–175
Junk removal, 303–304

K

Kelly, 123
Kick, 12, 38, 247–248, 312
    safety drill, 312
Kick detection equipment, 247–248
Kick/pit drill, 312
Kickoff point, 90–93, 97–98, 188
Kickoff tools/techniques, 90–93, 97–98, 174–177, 188
    jetting, 174–175
    whipstock, 175
    downhole motor with bent sub, 176
    downhole steerable motor, 176–177
Killing the well, 12

L

Land rig, 104–106
Lands, 76
Leaching control, 207
Lead time, 76
Limestone, 5–6
Lithology, 3–8
    shales, 4–5
    sandstones, 5
    carbonates, 5–6
    evaporites (salts), 6–8
Locating casing point, 95–96
Location preparation, 45
Logging, 43, 63–64, 70, 96–97
Logging program, 43

Index

Logging while drilling, 99, 232
Logistics, 277–279
Logistics coordinator, 264
Lost circulation, 212, 283–290
   surface hole, 284–287
   deeper formations, 287–288
   fractured/cavernous formations, 288
   diesel oil bentonite plug, 288–289
   cement curing, 289–290
Lost circulation curing, 212, 289–290
Luster, 217

M

Massive salt formations, 205–208
   salt–saturated slurry, 206–207
   setting times, 207
   leaching control, 207
   mud density, 207
Maximum allowable annular surface pressure, 60–61
Measurement while drilling, 93, 99
Mechanical skin, 75
Mechanical tools, 229–230
Medivac, 130
Metamorphic rock, 1–2
Metering specific targets, 271
Microresistivity tools, 226
Migration of oil/gas, 15–26
   primary migration, 19
   structural trap, 20
   reservoir rock, 20–21
   seal rock, 21–22
   secondary migration, 22–23
   reservoir drives, 23–25

Montmorillonite, 147–148
Mud classifications, 146–151
   dispersed mud, 147–148
   non–dispersed mud, 148–149
   solids free brines, 149–150
   oil mud, 150
   invert oil emulsion mud, 150
   air as circulating medium, 150
   aerated/foamed mud, 151
Mud density, 152–153
Mud design, 151–167
   physical properties, 152–153
   mud rheology, 154–164
   chemical properties, 164–167
Mud engineer, 265
Mud hydrostatic pressure, 140
Mud logging, 219–220
Mud pump, 114–115
Mud removal, 198–199
Mud rheology, 154–164
   Newtonian fluids, 156–157
   Bingham plastic fluids, 157–158
   pseudoplastic fluids, 158–160
   dilatent fluids, 160
   Herschel–Buckley fluids, 160–161
   time dependent rheology, 161
   flow regimes, 161–162
   design considerations, 162–163
   rheology model, 164
Mud solids content, 140–141
Mud system, 116–118
Multilateral wells, 183–184
Multiple wells from borehole, 173

N

Navigation (well), 187–188
Newcomers on rig, 310–311
Newtonian fluids, 156–157
Night drilling supervisor, 265
Night toolpusher, 267
Nipple up, 76
Non–dispersed mud, 148–149

O

Offset well, 30
Oil company/operator, 263
Oil base mud, 146–147, 150
Onshore to offshore reservoir, 173
Operations management, 263–281
    personnel, 263–268
    contract types, 268–270
    incentive schemes, 270–272
    decision making at wellsite, 272
    decision making in office, 272–273
    interfacing with service companies, 273–274
    estimating well cost, 274–278
    logistics, 277–279
    incident response, 279–280
Operations manager, 263–264
Operator (well), 74
Optimized drilling, 138–141
    rate of penetration, 140–141, 144
    mud hydrostatic pressure, 140
    mud solids content, 140–141
    drilling hydraulics, 141

Ordering equipment, 45–46
Orientated coring, 221–222
Over water work permit, 313–314

P–Q

Percentage of well cost saved, 271
Perforating, 72
Permeability (rock), 18–19
Permit to work systems, 313–314
    hot work, 313
    tank entry, 313
    over water work, 313–314
    electrical work, 314
Personal protection equipment, 316
Personnel, 263–268
Persons on board, 321
Petroleum composition, 16–17
Physical properties (mud), 152–153
    density, 152–153
    fluid loss, 153
    sand content, 153
Pipe body, 53
Pit volume totalizer, 117–118
Plate tectonics, 3
Polycrystalline diamond compact bit, 70, 135–137
Polymer, 167
Pore fluid sampling/analysis, 224–225
Pore pressure, 35
Pore space, 17–18
Porous media, 18
Pre–job meeting, 309

Index

Pre–spud meeting, 308
Pre–tour meeting, 309
Pressure coring, 222
Pressure (subsurface), 10–12
Primary migration, 19
Problems and solutions, 283–306
    lost circulation, 283–290
    stuck pipe, 290–298
    fishing, 298–306
Production buoy, 172
Production casing, 191–192
Production hole section, 68–69
Production liner running/cementing, 70
Production testing, 70–73, 232–235
    preparing for test string, 71
    running test string, 71–72
    perforating, 72
    well testing, 72–73
    well killing, 73
Proposal contents, 29–30
Prospect identification, 27–28
Pseudoplastic fluids, 158–160

R

Radio operator, 267
Radioactive sources, 305–306
Radioactivity tools, 229
Ram type blowout preventer, 241–244
    fixed pipe, 241–242
    variable bore pipe, 241–242
    blind, 243–244
Rate of penetration, 140–141, 144

mud hydrostatic pressure, 140
mud solids content, 140–141
Reactive shales, 165
Rebel tool, 180
Relief well, 174, 259
Remedial work/sidetrack, 173
Reservoir damage, 166
Reservoir drives, 23–25
    gas drive, 23–24
    water drive, 24
Reservoir fluids, 24–25
Reservoir rock, 20–21
Reservoirs (oil/gas), 15–26
    structural trap, 20
    reservoir rock, 20–21
    seal rock, 21–22
    reservoir drives, 23–25
Resistivity logs, 226
Resistivity/induction tools, 226
Returns, 76
Revolutions per minute, 139–140
Rheology (cement), 204–205
Rheology model (mud), 164, 167
Rig classifications, 104–113
Rig criteria, 103–104
Rig discharge/spill, 317–319
Rig moving, 46–49
Rig positioning, 40–41
Rig selection/equipment, 103–131
    suitable drilling rig, 103–104
    drilling rig classifications, 104–113
    rig systems/equipment, 113–129
Rig superintendent, 267

Rig systems/equipment, 113–129
Rock properties, 17–19
Rock strength/stress, 8–9
Rock types, 1–2
Roller cone bits, 133–135
Rotary drilling assembly, 100
Rotary equipment, 122–125
Rotary sidewall coring, 223–224
Rotary speed, 50
Rotary table, 47, 122–124
Roughneck, 267
Roundness, 218
Roustabout, 268
Running casing, 34, 309–310
    safety meeting, 309–310

S

Safety alerts, 314–315
Safety drills, 312–313
    fire, 312
    abandon rig, 312
    kick/pit drill, 312
    trip drill, 312–313
    hydrogen sulfide drill, 313
Safety equipment, 316
    personal protection, 316
    other, 316
Safety meetings, 307–310
    pre–spud, 308
    weekly, 308–309
    daily operations, 309
    pre–tour, 309

    pre–job, 309
    running casing, 309–310
Safety/environmental issues, 270, 307–321
    safety meetings, 307–310
    newcomers on rig, 310–311
    training/certification, 311–312
    drills, 312–313
    permit to work systems, 313–314
    safety alerts, 314–315
    equipment certification, 314–315
    safety equipment, 316
    STOP program, 316–317
    discharge/spill minimizing, 317–319
    environmental impact studies, 319
    severe weather, 320
Salt dome, 20, 173
    drilling, 173
Salt formations, 6–8, 205–208
Saltiness, 218
Salts (mud), 165
Salt–saturated slurry, 206–207
Sampling, 216–225
    surface, 216–220
    downhole, 220–225
Sand content (mud), 153
Sand control, 81
Sandstone, 5, 18
Seal rock, 21–22
Secondary cementing, 211–212
Secondary migration, 22–23
Secondary recovery, 24
Sedimentary rock, 1–2
Seismic survey, 75

Index

Self contained breathing apparatus, 321
Semi–submersible drilling rig, 106–107
Senior drilling engineer, 264
Sensator, 131
Setting times, 207
Severe weather, 320
Shales, 4–5
Shallow gas, 38–40, 75, 254–256
    dynamic kill, 254–256
    floating rigs, 256
Shear stress, 9
Show (hydrocarbon), 236
Single phase fluid, 23
Single surface location, 171–172
    tension leg platform, 172
    floating production and storage offshore, 172
    production buoy, 172
    jackup production platform, 172
Sleeve coring, 221
Slurry, 169
Snubbing unit, 129
Solids control equipment, 118–121
Solids free brines, 149–150
Solids sticking, 292–294
Sonic tools, 226, 229‹j‹å
Sorting, 218
Source rock, 16–17
Specific gravity (weight), 218–219
Sphericity, 218
Sponge coring, 221
Spudding the well, 48–52, 76, 86–87
Stabilized assemblies, 178–179

hold/tangent/locked assembly, 178
drop assembly, 178–179
build assembly, 179
Stabilizer, 61–63
Standpipe manifold, 115–116
Status of well, 44
Sticking mechanisms, 291
Stiffness, 188
Stimulation, 80, 236
Stimulation, 236
Stinger, 214
STOP safety program, 316–317
Strata, 8
Stratigraphic traps, 20
Structural trap, 20
Stuck pipe, 290–298
    sticking mechanisms, 291
    geometry related, 291–292
    solids related, 292–294
    differential sticking, 294–296
    jars/jarring, 296–298
Submersible, 113
Subsea systems, 246–247
Subsurface safety valve, 71–72
Support costs/overheads, 275
Surface casing, 38, 56–58, 190, 210–211
    cementing, 210–211
Surface hole, 38, 284–287
    lost circulation, 284–287
Surface location/positioning, 40–41
Surface sampling, 216–220
    mud logging, 219–220

Surge/swab pressures, 168–169
Surveying (well), 184–187
Suspending the well, 100
Swelling properties, 218

T

Tank entry permit, 313
Technical assistant, 264
Tectonically active areas, 3
Telescopic joint, 89–91
Temperature, 16, 196
Temperature log, 230
Temperature rating, 203–204
Template, 86–87
Tensile stress, 9
Tension, 192–193
Tension leg platform, 112, 172
Test string, 71–74
    preparation, 71
    running, 71–72
    removal, 73–74
Test stump, 281
Thickening time, 203
Time dependent costs, 275‹j‹å
Time dependent rheology, 161
Time–depth data, 40–41
Tolerance calculations, 35–36
Tool joint, 53
Toolface azimuth, 93
Toolpusher, 261, 267
Top drive, 124–125

Training/certification, 311–312
Transparency/translucency, 217
Transponder, 130–131
Trip drill, 312–313
Tripping, 77
True vertical depth, 94–95
Tungsten carbide insert, 134–135

U

Underbalanced drilling, 259–260
Undergauge, 144

V

Variable bore pipe blowout preventer, 241–242
Viscosity, 167

W–Z

Washover/basket tools, 301–303
Water base mud, 146
Water drive, 24
Water table, 24
Weathering, 1–2
Weekly meeting, 308–309
Weight on bit, 50, 139
Well completion, 30, 43, 83
Well control, 237–262
    primary, 237–238
    secondary, 238–239

Index

tertiary, 239
blowout preventer stack, 239–247
kick detection equipment, 247–248
well killing, 249–254
shallow gas, 254–256
special considerations, 256–260
Well cost estimation, 274–278
fixed costs, 274
time dependent costs, 275
depth dependent costs, 275
support costs/overheads, 275
contingency costs, 275
accuracy, 275, 277
Well design, 33–36, 43
Well killing, 73, 249–254, 256–257
Well objectives, 39
Well outflow configuration, 81
Well planning, 79–82
Well profile, 68–69
Well program, 36–44, 85
general information, 38
shallow gas, 38–40
well objectives, 39
potential hazards, 40
surface location/positioning, 40–41
general notes, 41
drilling notes, 41–42
drilling fluid design/maintenance, 42
wellbore trajectory, 42
casing design, 42
geological information, 43
logging program, 43
coring program, 43
well completion design, 43

well test information, 44
status of well, 44
Well proposal, 28–30
contents, 29–30
completion design, 30
Well safety, 74
Well surveying, 55
Well testing, 44, 72–73
Wellbore trajectory, 42
Wellhead, 72–73, 87–89
housing, 87–89
Wellpath control, 178–180
stabilized assemblies, 178–179
rebel tool, 180
Andergauge steering tool, 180
Baker Hughes Autotrak, 180
Wellsite drilling engineers, 265
Wellsite geologist, 265
Whipstock, 175
Wildcat well, 74
Wireline, 63
Wireline logging, 230–231, 304–305
Wireline/logging tools (fishing), 304–305
Workover hoists, 103, 129
Workover/working over, 129–130, 168